Getting Started in Electronic Troubleshooting

William E. Long

Reston Publishing Company, Inc.
A Prentice-Hall Company
Reston, Virginia

Library of Congress Cataloging in Publication Data

Long, William E.
 Getting started in electronic troubleshooting.

 Includes index.
 1. Electronic apparatus and appliances—
Maintenance and repair.   I. Title.
TK7870.L58   621.3815'028   78-14029
ISBN 0-8359-2487-4

© 1979 by Reston Publishing Company, Inc.
 A Prentice-Hall Company
 Reston, Virginia 22090

All rights reserved. No part of this book may be reproduced in any way, or by any means, without permission in writing from the publisher.

10 9 8 7 6 5 4 3 2 1

Printed in the United States of America

To Annette, for everything

Contents

Preface ix

Introduction Important Ideas Behind Electronic Circuit Troubleshooting 1
    I-1   Philosophy and Basic Assumptions 1
    I-2   Basic Electrical Concepts and the Organization of Circuits and Systems 8
    I-3   Energy Sources, Control Elements, and Loads 15
    I-4   Functional Subsystems within a Larger Unit; A Typical Example 21

Chapter 1   Troubleshooting DC Power Supplies 27
    1-1   Requirements of a Power Supply 27
    1-2   Equipment for Troubleshooting Power Supplies 28
    1-3   Basic Troubleshooting Procedures for Power Supplies 30
    1-4   Running Down Short Circuits 34
    1-5   Half-Wave Rectifier Power Supplies 37
    1-6   Full-Wave Rectifiers Using a Center-Tapped Transformer 40
    1-7   Bridge-Rectifier Power Supplies 43
    1-8   Half- and Full-Wave Voltage Doublers 44
    1-9   Voltage Regulation 45
    1-10   Ground Points and How to Locate Them 50

Chapter 2   Audio Amplifiers in Radios, Television Receivers, and Other Equipment 55
    2-1   Initial Considerations in Troubleshooting AF Amplifiers 56
    2-2   Signal Tracing and Signal Injection: Dynamic Checking of AF Amplifiers 63

2-3 Using Static (DC) Voltage Measurements as Trouble Clues   65
2-4 Push-Pull Amplifiers   70
2-5 Now, About Transistors   76

Chapter 3   Radio Frequency Amplifiers: Troubleshooting and Alignment of Intermediate Frequency Amplifiers in Radio and Television Receivers   **79**
3-1 Urgent Considerations in the Use of Instruments, Service Notes, and Procedures   80
3-2 Injecting Signals into RF Amplifiers: Practical Considerations in the Use of Signal Generators   82
3-3 Intermediate Frequency Amplifiers in Radio Receivers   83
3-4 Summary   99

Chapter 4   Radio Frequency Amplifiers: Preselectors, Mixers and Oscillators in Radio and Television Receivers   **101**
4-1 Troubleshooting the Front End of a Superheterodyne Receiver   102
4-2 Television Tuners   108

Chapter 5   Television Receivers   **113**
5-1 Similarities and Differences between Television and Radio Receivers   113
5-2 Functional Block Diagram of the Basic Television Receiver: Signal Paths and Trouble Symptoms   114
5-3 Key Checkpoints in Television Receivers: Narrowing Down the Problem   122
5-4 Locating Checkpoints and Other Practical Problems   133
5-5 Summary   137

Chapter 6   Synchronization and Automatic Gain Control in Television Receivers   **139**
6-1 How the Television Receiver Creates a Picture   140
6-2 Viewing the Horizontal Sync Pulse: A "Square Wave"   142
6-3 Separation of Sync Pulses from Video Information   144
6-4 The Vertical Sync Pulse   154
6-5 Automatic Gain Control in Television Receivers   155
6-6 Adjusting and Troubleshooting AGC Circuits   157

Chapter 7   The Vertical Sweep Circuit   **161**
  7-1   The Nature of the Vertical Sweep Circuit   162
  7-2   Vertical Deflection Circuit Troubles   164
  7-3   Special Procedures, Hints, and Difficulties   168
  7-4   Summary   172

Chapter 8   The Horizontal Sweep Circuit   **173**
  8-1   Troubleshooting the Horizontal Sweep System: Symptoms, Troubles, and Checks   174
  8-2   Picking a Starting Point: The Role of the Educated Guess   175
  8-3   Checking the High Voltage Power Supply   175
  8-4   Checking for Causes of Loss of Raster Unrelated to High Voltage: The Picture Tube and Electron Gun Voltages   178
  8-5   The Horizontal Amplifier: Functions and Failures   179
  8-6   The Damper Circuit   182
  8-7   The Horizontal Oscillator   184
  8-8   Horizontal Automatic Frequency Control (AFC)   185
  8-9   Summary   190

Chapter 9   Color Television: Elementary Servicing Procedures   **191**
  9-1   Similarities and Differences in Monochrome and Color Television Receivers   192
  9-2   Purity Adjustment in Three Gun Color Picture Tubes   194
  9-3   Convergence   199
  9-4   Problems in Color Television Receivers   201

Chapter 10   Troubleshooting Safety, General Procedures, and Special Problems   **211**
  10-1   Safety   212
  10-2   Following a Job Through   213
  10-3   Special Problems and Techniques   216
  10-4   Instruments, Tools and Miscellaneous Service Aids   218
  10-5   Parts and Inventory   221
  10-6   Troubleshooting Equipment Containing Integrated Circuits   221

Index   **225**

Preface

This book is written to provide a guide to basic practical troubleshooting of electronic circuits.

Troubleshooting of electronic circuits is one of the most frequent and important functions performed by practicing electronics technicians. The ability to systematically analyze circuits, locate faults, and repair them effectively is the most universal skill demanded of electronics technicians by employers. Troubleshooting skill is equally valuable to any hobbyist or home repairman.

Where does one learn or gain troubleshooting skill? Most textbooks used in schools concentrate on the *theory* of transistors, other devices, and circuits. Scarcely any textbooks discuss *practical troubleshooting*. Schools assume that the translation of theory into practice is a function of the laboratory courses that accompany the theory (lecture) courses.

Up to a point, one can agree with this position. But an examination of the organization and content of laboratory courses reveals that almost every basic course in DC, AC, and electronics uses the laboratory primarily to affirm the theory. That is, the student builds a circuit, makes measurements and observations, acquires data, and then compares his data with the theoretical expectancy to affirm that the theory does indeed carry over into practice. Seldom does a student troubleshoot a nonworking circuit which someone else has built until after he has had a great deal of theory.

The truth is that laboratory work is only a first step in learning how circuits work and how to repair them when they fail to work properly. Really learning how to apply theory to practical troubleshooting does not often occur until on-the-job experience can provide the missing elements. There is a great need for a sound and easy-to-follow guide to aid people to "get started" and to acquire practical troubleshooting know-how more quickly and easily than is now possible. The need exists whether one is in school or out, a hobbyist, a home repairman, or an aspiring electronics technician.

The introduction of this book discusses the foundations and philosophy of electrical and electronic troubleshooting, the basic concepts and assumptions, and the organization of circuits. Chapter 1 deals with techniques of isolating troubles in power supplies and the use of instruments. Together, the introduction and first chapter present and discuss an underlying core of very basic information that is repeatedly called upon later when analyzing other circuits in detail.

The organization of the book is flexible; the sequence of use can be determined by the needs of the user. Each chapter deals with a particular aspect of troubleshooting and is intended to stand by itself as a unit.

In seeking the most valuable examples to help the greatest number of people, it seemed logical that the most readily available basic circuits would be the best choice. What could be a better choice than equipment normally found in the home? Thus, circuits typical of radio receivers, audio amplifiers, and television receivers form the basis for most of the discussions found in this book.

The choice of home equipment for our examples in no way weakens the material's effectiveness for nonhome type equipment, since basic strategy and troubleshooting procedures are identical. Actually, the wide availability of the examples used should enhance learning and serve as a bonus to most of us who will most certainly face at some time a stereo set that distorts the sound, a radio with an excessive hum, or a television receiver with a black line at the bottom, top, or sides of the picture.

Troubleshooting should always start with a careful, critical, and thoughtful look at the *symptoms* displayed. The reason is that symptoms are the visible portion of the cause-and-effect duo, an association that the effective troubleshooter uses to full advantage. Symptoms provide timesaving clues that frequently point to the particular subsystem where the fault lies, or at least suggest a starting point for making further checks and instrument measurements. The following troubleshooting guides to the material in this book are included for easy reference.

Symptom	Chapter
No sound in record players and other AF amplifiers	Chapters 1 and 2
No sound in AM and FM radios	Chapters 1, 2, 3, 4
No sound in monochrome and color TV	Chapters 2, 3, 4, and 5 (Table 5-1 Figure 5-1)
No picture in monochrome and color TV	Chapter 5 (Table 5-1 and Figure 5-1)
No brightness in monochrome and color TV	Chapter 5 (Table 5-1 and Figure 5-1)
Sweep or sync problems in monochrome or color TV	Chapters 5, 6, 7, 8, and 9
Loss of color, color wrong or impure, color not in right place (Convergence) in color TV	Chapter 9
Safety and special problems	Chapter 10

Topic	Table or Figure
Power supplies	Table 1-1
Audio amplifiers	Table 2-1
RF and IF amplifiers	Table 3-1
General TV troubles	Table 5-1
TV sync problems	Table 6-1
TV AGC problems	Table 6-2
Vertical linearity problems	Table 7-1
Vertical frequency problems	Table 7-2
Horizontal sweep circuit problems	Table 8-1
General color TV problems	Table 9-1
Convergence sequence	Figure 9-5
Color TV problems	Table 9-2
FCC color TV signal standards	Table 9-3
General troubleshooting flow chart	Table 10-1

Good luck and productive troubleshooting!

introduction

Important ideas behind electronic circuit troubleshooting

We live in the Age of the Computer, with stereo, FM, AM, MASERs, LASERs, BJTs, FETs, MOSFETs, SCRs, ICs, OP AMPs, and LSI, to mention only a few. With electronics achievements as the enabling factor, we have gone to the moon and returned safely. Individually and collectively, we are surrounded by and depend upon electrical and electronic devices to an extent that few of us fully comprehend.

The twentieth century will undoubtedly be known to future historians as the dawning of the Electrical Age. It is in this century that alternating current has grown to be a dominant energy distribution system in the world, largely because it represents the most flexible and economical system that man has yet devised. Wireless communication, which is electrical and electronic in nature, is a significant twentieth-century development because it represents the ultimate in speed—since radio waves travel at the same velocity as light, 186,000 miles per second.

The vast number and variety of electrically operated equipment in use today makes maintenance an urgent problem. Anyone who knows how to repair malfunctioning electrical and electronic equipment is presented with both a challenge and an opportunity.

I-1 Philosophy and Basic Assumptions

I-1a The Role of Troubleshooting in a Recycling Society

Our world is presently stirring with a new awareness that the earth has limits: There is only so much fossil fuel, farmable land, drinkable water, ability to absorb garbage, and potential to support life. We now perceive that the time is rapidly approaching when it will be imperative that we recycle metals, pulpwood, water, and resources of all kinds because nature's storehouse of easily acquired raw materials will be empty.

In a sense, man has always employed recycling in his societies. Every article that has been repaired and restored to normal function rather than discarded has undergone one kind of recycling. Every farmer in history who has fertilized his fields with animal droppings to produce better crops has engaged in another kind of recycling. And ever since man invented language he has been recycling information!

What, then, is the role of troubleshooting in a highly developed, recycling society?

For the most part, we who live today are accustomed to a largely linear way of using resources (see Figure I-1a). We process raw materials into devices that have planned obsolescence to encourage us to discard them quickly and periodically purchase new ones made of fresh raw materials. Is it any wonder that our garbage dumps are monumental? It seems highly possible that our descendents will mine our garbage dumps for the raw materials that we throw away today. Our grandchildren will live in a society that operates much more in a recycling mode (see Figure I-1b) out of a necessity created by long-standing ways of doing things.

Any product or machine has an economic life that begins when the object is first placed in service. This useful economic life ends at the point when the cost of repair and maintenance becomes prohibitive, or when the cost of continued use is greater than the cost of replacement after all factors are considered. The point at which an item is discarded *must* advance in an age of scarcity, and the item must be repaired many more times than is present practice (see Figure I-2). The prospects we face in the future will place increasing emphasis upon all aspects of maintenance, with equipment repair a vital part. Whether or not electrical and electronics repairmen will ever approach the status presently enjoyed by those who repair and maintain the human body — doctors and dentists — remains to be seen.

I-1b The Service Orientation: Basic Assumptions and Exceptions

The electronics troubleshooter or technician, like other experts, acts upon certain basic assumptions:

1. *The troubleshooter assumes that the design work on the equipment has already been done.* Hence, the troubleshooter

is looking for some part, connection, or adjustment that has broken down or changed in value; he is not trying to determine what the best values might be for parts or circuits, because this is the design function which he assumes has been done.

2. *The troubleshooter assumes that the equipment was once working properly, and that the circuit, the wiring, and the connections are or were correct.* His problem is to isolate the part that has changed, restore it to its original condition, and then check for proper operation.
3. *The troubleshooter assumes that he is looking for a single trouble in most cases, not widely dispersed, unrelated, multiple troubles.*
4. *The troubleshooter assumes that electrical and electronic equipment will behave according to theory.* If it seems not to, it is because he is overlooking something.
5. *The troubleshooter assumes that his time is important and valuable.* Thus, as he works, he is constantly weighing his choice of instruments, sequence of procedures, availability of parts, and adequacy of service information to save time.

Of course, there are exceptions to these basic assumptions that apply in the following instances: if equipment has been tampered with or modified; if equipment has never worked properly; or if equipment has just been built.

In dealing with these exceptions the troubleshooter can not assume that the wiring connections and parts values are all correct, that there is probably only a single trouble, or that adjustments are anywhere near proper. In these situations the troubleshooter's task is vastly more complicated, for he must check everything and can take nothing for granted.

The service technician deals almost exclusively with equipment that someone else has constructed. This is in contrast to the R & D (research and development) technician who frequently builds and modifies circuits and prototypes, accumulates data, and is much concerned with determining the appropriate component values and circuits to use.

I-1c Proper Static Conditions: A Prerequisite for Dynamic Performance

In all troubleshooting, there is a rank of importance attached to information, circuit conditions, and steps in repair procedures.

4 Introduction

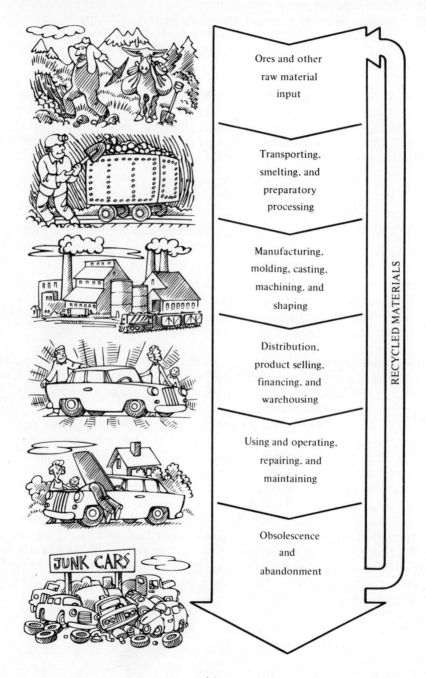

(a)

I-1 Philosophy and Basic Assumptions 5

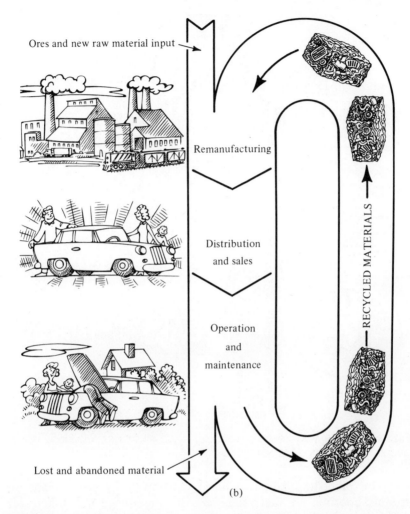

Figure I-1 (a) Present industrial societies operate largely in a linear fashion that requires constant input of fresh raw materials at the beginning and ends with abandonment of goods produced. (b) As new raw materials become scarce, recycled materials must provide more of the input to the manufacturing process.

This must be clearly recognized and taken into account in order to minimize lost time.

1. Proper DC static conditions must first exist in order for a circuit to perform its dynamic signal-handling function properly.
2. Final pinpointing of the specific cause of a malfunction is accomplished in the majority of cases by a check of static conditions.

Figure I-2 Are **you** keeping your _____ (fill in car, clothing, appliances, furniture, etc.) longer?

The preceding facts are used constantly by technicians in troubleshooting. In fact, they are so ingrained and taken for granted by skilled technicians that they are often overlooked when instructing others. Like driving on the right side of the road in the United States, these facts are so generally understood by professionals in the field that they sometimes forget that the less skilled do not know them.

A logical question that might well be asked at this point is: What constitutes proper static conditions in electronic circuits? These factors are:

1. All component values are what they should be.
2. All connections and wiring are correct.

3. All DC (static) voltage and current levels at all points in the circuit are the correct values when no signal is being processed by the equipment.

I-1d Experimental Data: The Final Authority on Circuit Performance

Experimental data refer to measurements and observations of circuit characteristics and performance *taken from an actual working circuit*. As long as the instruments used to make measurements yield reliable data, the data will reflect the true condition of the equipment. Reliable data are, therefore, the final authority on circuit performance, and such data are always experimentally derived.

Using experimental data can be better than using theory. While calculations based on theory and equations which show a one-to-one relationship with circuit characteristics are invaluable as predictors of circuit performance, there are many ways things can go wrong. For example, the wrong equation may be used, an error may creep into the figures or the decimal point may be misplaced and, very often, minor factors are ignored in the equations to simplify calculations. Furthermore, the values of components that emerge from calculations are usually shifted to the nearest standard value to keep costs down. There are some values, such as capacitances and coupling between circuit conductors, which are impossible to know exactly until the circuit is actually built. These values may produce unanticipated effects. Actual construction of a circuit adds to the list of things that can go wrong. Parts values may occasionally be outside of tolerance, poor or wrong connections may be made inadvertently, very delicate parts may be damaged by heat, voltage, or stress during the construction, and certain parts may be put in backwards in a careless moment.

Experimental data are essential in optimizing values, adjusting and tuning electronic equipment. This is the special province of the electronics technician at the manufacturing, operating, and maintenance levels. Experimental data are, moreover, the only way to determine if the performance of the electronic equipment meets design specifications.

Many circuits include provisions for adjustment to bring the circuit to optimum performance. In virtually every case, setting the adjustment to the right point requires the use of experimental data to tell when the right point has been reached.

I-2 Basic Electrical Concepts and the Organization of Circuits and Systems

All energy systems have certain concepts in common. An electrical system is often compared to a closed hydraulic (fluid) system to illustrate parallel ideas. In an electrical system, pressure becomes voltage; pipes and hoses become wires or conductors; the entire system becomes a circuit; gallons per minute become coulombs per second (amperes); drag or friction becomes resistance; and valves become switches. The amount of flow in both systems depends on how much pressure or voltage is applied to how much resistance.

I-2a Voltage, Current, Resistance, and Work

Figure I-3 illustrates a closed hydraulic system and an electrical system in an automobile. Note that both require input energy (pressure) from a pump, battery, generator, or other source to make the systems work. The point is that all useful electrical and electronic systems require a voltage source in order to operate. When the voltage source is connected to a closed circuit (the switch is "on"), the amount of current that flows depends on the voltage and the resistance.

Current is a movement of charged particles, normally electrons, that results when voltage is applied to a closed circuit. Direct current (DC) circuits are those in which the direction of current does not change, even though the amount of current may vary up and down. Direct current circuits begin with a source of DC voltage, such as a battery, DC generator, solar cell, or thermocouple, to name some of the best known. Rectifier-filter types of power supplies change AC input to DC output and act as battery substitutes in most electronic equipment.

Any time current flows in a resistor, or some device that changes electrical energy into another form, by definition, work is done. Generally, we are most interested in the *rate* of doing work, termed *power*. Electrical power is measured in watts. One horsepower is equivalent to 746 watts.

I-2b Open and Closed Circuits; Series and Parallel Circuits

Just as there are certain concepts that are fundamental to all electrical circuits, there are certain basic ways that all electrical circuits are organized and put together. A troubleshooter should un-

I-2 Basic Electrical Concepts 9

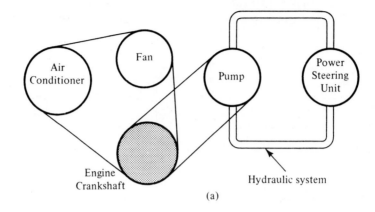

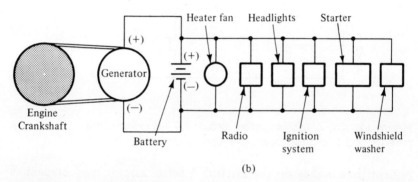

Figure I-3 Energy systems in an automobile. (a) Engine burns gasoline to turn crankshaft which furnishes energy input to drive air conditioner, fan, power steering, and electrical generator. (b) Generator charges battery, which stores energy to operate the various parts of the electrical system.

derstand these and keep them in mind if he hopes to use his time effectively. We shall briefly review here major topics in circuit organization.

A *circuit* is any path through which electric current can flow. If the path or circuit is not interrupted or broken, it is referred to as a *closed*, or *complete circuit*. If interrupted or broken, the path is referred to as an *open*, or *incomplete circuit*. A switch is a control device that enables us to open and close (break and complete) a circuit conveniently at will. Refer to Figure I-4 for details.

A *series circuit* is one in which there is only a single path; therefore, every bit of current must pass through every com-

10   Introduction

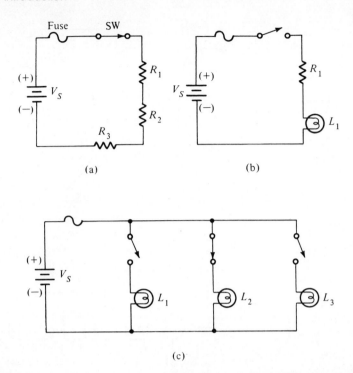

Figure I-4 (a) A closed series circuit. (b) An open series circuit. (c) A parallel circuit with two open branches and one closed branch.

ponent that makes up the circuit. A series circuit may consist of many, or very few, components, provided that they form only a single path for current. A basic law of a series circuit is that current is the same value *at all points*. Hence, if current must be measured in a series circuit, one should always choose the most convenient point, since the reading will be the same no matter where it is taken.

A *parallel circuit* is one in which there is *more than one path* for current. There can be many paths (more than one) for current, as shown in Figure I-4c. A basic law of parallel circuits is that the voltage is the same across all the branches. Hence, when measuring voltage in a parallel circuit one should always choose the most convenient point, since the reading will be the same no matter which branch is measured.

It is reassuring to realize that there are only these two basic kinds of circuits, series and parallel, when they are classified according to the arrangement of their parts. Combinations of series circuits and parallel circuits are designated *series-parallel circuits*.

These can be much more complex when many components are involved than either basic series or parallel circuits alone.

A major obstacle in dealing with electricity is that the most important events take place on the atomic scale, which is much too small to be visible to the human eye. We cannot tell by looking at a circuit if there is current or not, or if there is voltage or not. Nor can we evaluate any of its other electrical characteristics. We must use *instruments* (which respond to electrical characteristics) as translators of the desired electrical characteristic into a form that is meaningful to human senses.

Thus we have ammeters to measure current, voltmeters to measure voltage and ohmmeters to measure resistance. Usually all three electrical characteristics can be measured by a single multipurpose instrument constructed around a basic milliammeter or microammeter. The instruments which have these capabilities are known as a *VOM* (*V*olt-*O*hm-*M*illiammeter), a *VTVM* (*V*acuum-*T*ube-*V*olt-*M*eter) or, more recently, a *FETVOM* (*F*ield-*E*ffect-*T*ransistor-*V*olt-*O*hm-*M*eter). Each of these instruments has distinct characteristics which may make one more desirable for certain measurements under particular conditions than the others. This aspect of selecting the best instrument for a particular measurement will be considered as it applies to each topic under discussion.

It should be clearly understood that voltage and resistance measurements form the backbone of basic troubleshooting procedures. Current measurements take more time and are usually avoided when possible.

I-2c Mathematical Predictability, Ohm's Law, and Electron Flow

Ohm's Law is one of the most fundamental and useful rules that govern circuit behavior. It is a simple mathematical model. According to Ohm's Law, the current in a circuit is directly proportional to the voltage or pressure and inversely proportional to the resistance. The formula is stated as:

$$I = \frac{V}{R}$$

or

$$R = \frac{V}{I}$$

or

$$V = IR$$

In the formula, *I* is in amperes, *R* is in ohms, and *V* is in volts.

Through the use of Ohm's Law, we can calculate *I*, *V*, or *R* if two of the three are known. This is one way to avoid making a current measurement: measure voltage and resistance, then calculate current. This will usually take less time than actually measuring the current, as we mentioned earlier. Ohm's Law applies to both AC and DC circuits.

Ohm's Law illustrates a very important point about electricity and electronics which we might term *mathematical predictability*. Because of this, we can assume any convenient value of voltage and then calculate the amount of resistance needed to permit a desired value of current. In other words, we can find the value of current without having to build the circuit and make actual measurements! Ohm's Law shows a one-to-one relationship with the actual circuit, which means that important characteristics of a circuit can be found on paper without having to construct the circuit unless confirmation is desired. In fact, almost the entire basis of design is founded on the one-to-one relationship that various mathematical equations have with real processes and hardware. Errors and discrepancies creep in because practical manufacturing today does not permit component values of consistent precision to match exactly the mathematical analogue. So when the circuit is actually constructed the parts values used are seldom, if ever, exactly the same as the values used in the design calculations.

There are two basic types of current, alternating current (AC) and direct current (DC). As stated earlier, in direct current circuits the charged particles always move in the same direction. It is generally agreed that electrons (negative charges) move from the negative terminal of a battery through the external circuit to the positive terminal. That is, *electrons move from negative to positive*. Here, we have an unfortunate controversy. Before it was convincingly shown that electrons move from negative to positive, it was thought that whatever moved was moving from positive to negative. Current regarded as moving from positive to negative is known as *conventional current*. Most textbooks and general practice today continue to use conventional current. An exception is found in electronics technician training in our military, which considers electron flow as synonymous with current and teaches that current moves from negative to positive! Other than the confusion resulting from the use of dual terminology, little harm is done since all formulas yield valid results.

Alternating current may be thought of as current in which the charged particles move back and forth, much like a pendulum. AC is also sometimes described as current that constantly varies in amplitude (amount) and periodically reverses in direction.

I-2d Resistance, Inductance, and Capacitance: The Three Basic Attributes of All Circuits; Components as "Lumps" of R, C and L

Resistance *(R)*, inductance *(L)*, and capacitance *(C)* are the three electrical characteristics in all electrical and electronic circuits. Resistance changes electrical energy into heat and, if heat is high enough, into light. Inductance and capacitance store energy in magnetic and electric fields and later return the stored energy to the circuit.

The rate at which resistance changes DC and AC electrical energy into heat we call power:

$$P = V_R I_R \text{ watts}$$

or
$$P = I^2 R \text{ watts}$$

or
$$P = \frac{V^2}{R} \text{ watts}$$

Unlike resistance, inductance and capacitance act to oppose current in a way that is *dependent upon frequency*. This opposition to current is called reactance *(X)* and is primarily associated with AC.

$$\text{For inductance, } X_L = 2\pi f L \text{ ohms, or } X_L = \frac{v_L}{i_L} \text{ ohms}$$

$$\text{For capacitance, } X_C = \frac{1}{2\pi f C} \text{ ohms, or } X_C = \frac{v_C}{i_C} \text{ ohms}$$

Note that AC voltages and currents are stated in lower case letters *(v, i)*.

Resistors, inductors, and capacitors may be thought of as "lumps" or concentrations of a specific amount of *R, L* or *C*. A circuit may act predominantly like one of these, depending on the comparative values of the other two. Circuits are often classified as resistive, inductive, or capacitive on this basis, even though a straight piece of wire by itself possesses all three characteristics to some degree.

Each type of component *(R, C or L)* is different from the others in its effect on a circuit. Consequently we use a resistor, capacitor, inductor, or some combination of these to accomplish different things. For example, a resistor may be used to do the following:

1. Limit the current to a desired value $\left(I=\dfrac{V}{R}\right)$, as in a motor controller, instrument panel lighting in an automobile, or electronic circuit.

2. Establish a certain voltage across it $(V = IR)$, as in electronic circuits to produce the proper static conditions that permit transistors and vacuum tubes to amplify signals as desired.

3. Produce a certain amount of heat (Power in watts $= I^2R$; $P=\dfrac{V^2}{R}$; or, $P = VI$), as in a soldering iron, an electric stove, an electric blanket, and an electric coffeemaker.

As mentioned earlier, a resistor changes electrical energy into heat in accordance with the above formulas. It does not store electrical energy. On the other hand, capacitors and inductors do store small amounts of energy temporarily. Their stored energy is later returned to the circuit. Theoretically, capacitors and inductors do not change electrical energy into heat as does a resistor; hence, they have less tendency to heat up in a circuit.

There are two kinds of fields in electric circuits. A *field* is the mechanism by which charged particles make themselves felt. The two kinds of fields are termed *magnetic* or *electromagnetic* and *electric* or *electrostatic*. An electric field is associated with stationary charges, whereas a magnetic field is associated with moving charges. A *capacitor* is an electric field device or component, and may be thought of as a "lump" or concentration of the electric field. An *inductor* (coil) is a magnetic field device and may be thought of as a concentration or "lump" of magnetic field.

A capacitor may be used in a circuit to accomplish the following:

1. Store energy and act as a small reservoir of current and voltage, as in a filter in a power supply.

2. Act as an open circuit for DC and a closed circuit for AC, as in a coupling capacitor in an audio amplifier.

3. Cancel the effects of inductance, as in tuning (resonant) circuits.

An inductor, sometimes referred to as a choke, may be used in a circuit to accomplish the following:
1. Store energy and act as a small reservoir of current and voltage, as in a filter, with unique abilities to produce higher pulse voltages than the original source voltage when the stored energy is returned or "dumped" to the circuit rapidly, as in the coil of an automobile ignition system.
2. Cancel the effects of capacitance, as in tuning (resonant) circuits.

In working circuits, troubleshooting of capacitors and inductors involves measuring their resistance and DC and AC voltages across them in much the same way as for resistive circuits.

I-3 Energy Sources, Control Elements, and Loads

It may seem obvious that all useful electrical and electronics circuits have in common energy sources, control elements, and loads. It is stated here to emphasize strongly this fact of system organization. *If a malfunction occurs in any part of the circuit, proper operation of the entire system is jeopardized.*

I-3a Power Distribution Systems Compared to Electronics Systems

It is useful to make a distinction between power distribution systems and most electronics systems. Most commercial power distribution systems that cover large areas are alternating current systems. Feasibility and cost factors give AC a huge advantage over DC in large-area use. DC is often used in small-area power distribution (and electronics) systems such as automobiles, submarines, and other applications where energy must be stored. Note that these systems are portable and comparatively small. But the real advantage of DC is that it can be stored in batteries. AC is at a great disadvantage because it cannot be stored, but must be used as it is generated.

A special feature of power distribution systems is their heavy use of parallel circuits, which permits loads to be manufactured for standard voltages, such as household appliances. Another great advantage of parallel circuits in such use is the fact that each appliance or load may be turned on and off independently, without affecting other loads.

Electronics systems do not favor parallel circuits as overwhelmingly as do power distributions systems. Rather, they include a *combination* of series circuits, parallel circuits, and series-parallel circuits.

On the other hand, electronics systems typically combine AC and DC to accomplish their function. First, AC power from a commercial power system is brought in and changed to DC, which is required for proper operation of transistors and vacuum tubes. Specific DC levels of voltage are established at device terminals to fulfill the essential static conditions. Subsequently, when a signal is processed in the circuit, variations are set up in these previously established DC conditions in accordance with the signal. Such dynamic conditions are variously described as *DC with an AC component,* varying DC, and in some cases, pulsating DC.

The heart of electronics systems is composed of amplifying devices such as transistors and vacuum tubes, plus diodes and a growing number of other devices not essential to power distribution systems. A very large percentage of the devices and components in a typical electronics system operates at power levels of less than one watt. In fact, voltage and current levels in microvolts, millivolts, microamperes, or milliamperes are dealt with by electronics technicians every day. To a person accustomed to working with a power distribution system, the amounts of power dealt with in electronics seem incredibly small, and he may find it difficult to believe they account for the results he sees all around him.

Summing up, power distribution systems exist to distribute energy as economically as possible to various physical locations where it will be used to do work. Electronics systems exist primarily for the purpose of amplifying and processing time-varying signals to accomplish a very wide variety of human ends.

I-3b Energy Sources, Control Elements, and Loads in Working Circuits

Electrical energy sources almost always act as translators of energy from some other form. Present commercial AC power distribution systems are powered by alternators which are large machines that translate mechanical energy into electrical energy. The mechanical energy may be supplied by falling water, steam, or any engine or source capable of turning the alternator. Some of the more common translations in use today are shown below:

I-3 Energy Sources, Control Elements, and Loads

Energy Input	Translator or Transducer	Output
Mechanical	Alternator, generator	AC, DC voltages
Chemical	Batteries, fuel cells	DC voltage
Heat	Thermocouple	DC voltage
Light	Solar cell	DC voltage

Whatever the basic source may be, the electrical energy it produces is sent to the controlled circuit which in turn delivers its output to the desired load (see Figure I-5). The vast majority of electronic equipment in the United States is designed to operate from a large commercial AC power distribution system which is standardized at 117 volts, 60 Hertz (cycles per second). Higher or lower voltages are readily produced by a suitable transformer if needed.

Control elements are those components, devices, and circuits which control the flow of electrical power in a way that satisfies the equipment user's objectives. Switches, relays, and controllers of varied complexity are the major control elements in power distribution systems. In electronics systems, in addition to switches and relays, we find transistors, vacuum tubes, diodes, and a growing number of items peculiar to electronics, all of which control voltage and/or current in some way.

The organization of a typical electronics system is shown in Figure I-6. One can quickly see that an electronics system differs from a power distribution system in a number of ways:

1. The number one priority in an electronics system is proper handling of the signal information.
2. Amplifiers require DC voltage for satisfactory operation.
3. Electronics systems have two energy inputs, one small (the signal source) and one large (the DC power supply). The small energy input comes from the input transducer and acts to control the flow of energy from the main power supply through the transistors and other amplifying devices.

We shall be referring to the organization depicted in Figure I-6 repeatedly in discussing troubleshooting. It is so useful that it should be securely incorporated in a troubleshooter's working knowledge and become second nature.

A transducer, as explained earlier, is a device which changes energy from one form to another. Note that the typical electron-

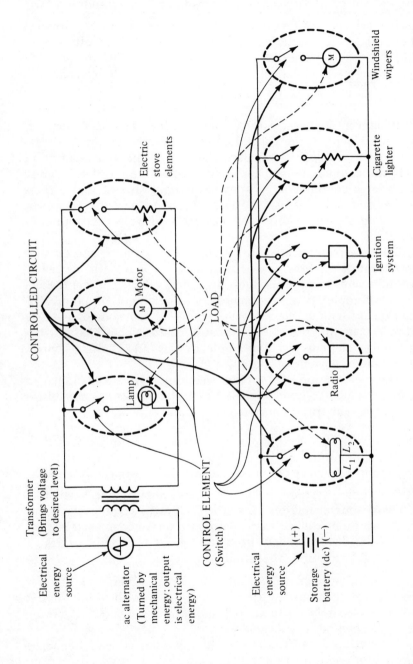

Figure I-5 Two power distribution systems, one AC and one DC, showing energy sources, controlled circuits and loads.

I-3 Energy Sources, Control Elements, and Loads 19

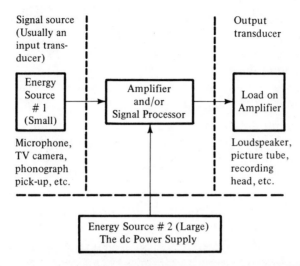

Figure I-6 Typical organization of an electronics system consisting of an input transducer, an amplifier or signal processor, an output transducer, and a DC power supply to provide the power required by the amplifier.

ics system incorporates an input transducer (to produce the electrical signal) and an output transducer (to translate the electrical output signal back into the form of energy desired). The output transducer acts as the load on the electronics system.

To summarize, we observe that electronics equipment is made to amplify or process in some way a time-varying signal voltage or current. Ordinarily this signal comes from an input transducer which changes some characteristic of its environment (vibration, light, heat, pressure, etc.) to an electrical impulse. After being amplified to the desired power level the signal is connected to "drive" an output transducer such as a loudspeaker, a picture tube (CRT), or other device which converts the amplified signal back into the desired form.

I-3c The Troubleshooter and "What Should Be There"

It is imperative that the electronics troubleshooter know "what should be there" if he is to work effectively. In addition to knowing the general organization of electronics equipment previously discussed, he must know circuit details precisely.

A list follows of what the troubleshooter must have or know:

1. *He needs schematic drawings showing component values and connections.* Would you attempt to drive your car to

Franklin, Pennsylvania *without* a road map? How long would it take?

2. *He needs to know DC (static) voltages to be found at strategic points when the circuit is working properly.* How can he know if a measured value is right or wrong if he doesn't know what it should be?
3. *In many circuits he needs to know AC signal voltage levels and waveforms at strategic points when the circuit is working properly.* Again, how can he know if a measured value is right or wrong if he doesn't know what it should be?
4. *He needs to know any special handling precautions, adjustments to be made, or other information peculiar to this particular piece of equipment.* How can he be sure he will not create more problems if he makes a mistake?
5. *He needs to know part numbers, manufacturers, and local sources of components found to be defective.* He does want to repair the unit, doesn't he?
6. *He needs to know specifications of the unit, the input signal level, the output signal level, and frequency characteristics.* How will he ever know if the equipment is performing normally if he doesn't know what "normal" is?

We have emphasized that it is imperative that the electronics troubleshooter know "what should be there" if he is to work effectively. There are three major avenues by which such information is available to him. They are:

1. Out of his own head, that is, from training, experience, and a good memory.
2. By asking someone who knows, that is, depending on someone else with training, experience, and a good memory.
3. From manufacturers' specifications, literature and service notes, that is, building and keeping a library.

There are, in addition, other stratagems that are less practical:

1. Comparing all measured values of the malfunctioning unit with values obtained from a correctly operating unit. This is a good procedure, but impractical because an identical unit is seldom available.
2. Seeking to correct the trouble by experimentally substituting known good parts for the suspected bad parts and then check-

ing for proper operation. This is *not* a recommended procedure unless the troubleshooter has some notion of what should be there, because the risk of introducing new and greater trouble is very high.

I-4 Functional Subsystems within a Larger Unit; A Typical Example

Almost all electronics systems of any appreciable size include a number of subsystems. These subsystems generate signals, handle other frequencies, or serve a purpose different in some way from remaining parts of the circuit. A small, ordinary car or home radio receiver will illustrate our point and serve to make clear a few more basic considerations and sequences involved in troubleshooting.

Figure I-7 shows the major functional blocks or subsystems in our small radio receiver. These include the following:

1. The DC power supply.
2. The audio frequency (AF) amplifier.
3. The intermediate frequency (IF) amplifier.
4. The radio frequency (RF) mixer-converter.
5. The local RF oscillator.

Note that each of these handles different frequencies (AF, IF, RF), and some serve different purposes. That is, the oscillator generates an RF signal, whereas the mixer-converter accepts one RF signal from the antenna and another from the RF oscillator and from them produces the IF signal.

How does this fit in with the general organization of electronics systems shown in Figure I-6? Is it different? Does it change the general organization? Let us compare Figures I-6 and I-7. Both have a DC power supply, input signals (from antenna and microphone), and output signals (to speakers). But in Figure I-7 the signal processor-amplifier consists of a number of subsystems. Figure I-7, in other words, is in slightly more detail than Figure I-6. Otherwise, the ideas presented are identical.

Detailed troubleshooting of a small radio will be discussed in the following chapters as a follow-up to lend continuity to the ideas we have presented here in the Introduction. There are, however, some general ideas in troubleshooting which are appropriate to introduce at this point:

22 Introduction

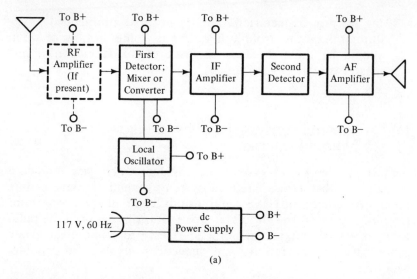

Figure I-7 (a) Block diagram showing similar functional subsystems in two radios. (b) Schematic of a typical low-cost table model radio, tube type (courtesy of Howard W. Sams & Co., Inc.). (c) Schematic of a low-cost car radio using transistors rather than vacuum tubes (courtesy of Howard W. Sams & Co., Inc.). Note that the car receiver has an RF amplifier stage dotted in); this is usually found in more expensive, better quality receivers.

1. *All* of the subsystems of the radio will perform unsatisfactorily or not at all if the DC power supply output is improper.
2. Failure of *any* subsystem to function properly will show up in this case as a failure of the total system to operate correctly. This is how the user discovers that something is wrong.
3. Dynamic operational checks of each subsystem to identify the faulty one are a good way to save time. This is generally referred to as *localizing* the trouble so that a minimum amount of time is spent checking the good subsystems for the defect.

It should be noted here that dynamic operational checks (Item 3, above) require that each subsystem be tested and the following questions answered:

1. With a proper signal input, does the subsystem produce the proper signal output?

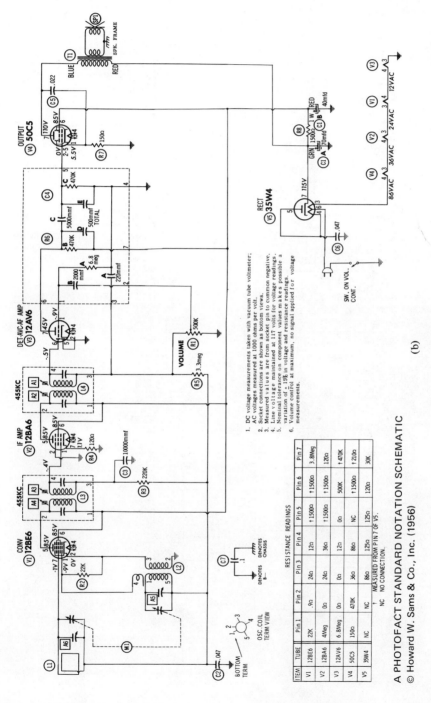

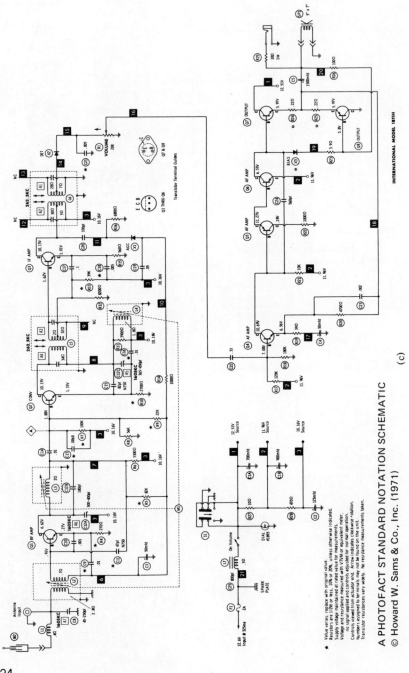

2. After dynamic checks reveal the faulty subsystem, what improper static condition is the cause?
   a. Do bias and other DC voltages at tube or transistor terminals agree with what should be there?
   b. Are all connections good?
   c. Have any components changed in value?

This concludes our introduction to important ideas behind electronic circuit troubleshooting. In Chapter 1 which follows, we examine details of how to go about troubleshooting on one of the functional blocks in an electronics system, the DC power supply.

chapter 1

Troubleshooting DC power supplies

No electronic system will operate satisfactorily without an adequate power supply. For nearly all systems, this means a DC power supply. Batteries are sometimes used to supply DC, but they are costly and must be recharged or replaced frequently. Hence, a rectifier-filter type of power supply that operates from commercial AC power lines is preferred and found in most equipment. Such power supplies act as battery substitutes.

The DC power supply is one of the first areas a technician checks when faced with a malfunctioning piece of electronic equipment, since the power supply must work in order for the equipment to work. What does the troubleshooter look for?

1-1 Requirements of a Power Supply

Any type of power supply must provide the following for the equipment it serves:

1. The required level of DC voltage and current.
2. An adequately filtered and regulated DC output.
3. For vacuum tube equipment, the proper level of AC or DC voltage to heat filaments and heaters.

When trouble develops in a power supply, it shows up as a failure to fulfill one or more of these functions. That is, DC voltage output might be too low, or inadequately filtered or regulated, or tubes might not heat properly.

Rectifier-filter power supplies that operate from commercial AC power lines normally include a transformer, one or more rectifiers and a filter.

1-2 Equipment for Troubleshooting Power Supplies

Troubleshooting of basic power supplies requires only standard test equipment that need not be of highest quality as long as it is dependable. The following is a list of equipment basic to troubleshooting power supplies in home, industry, or laboratory:

1. *Required*
   DC and AC volt-ohmmeter: VOM, VTVM, or FETVOM
2. *Highly recommended*
   Isolation transformer with variable output
   Oscilloscope
3. *Very helpful*
   Capacitor checker
   DC milliammeter
   Tube checker
   Transistor and solid state diode checker
   Wattmeter
4. *Helpful but not essential*
   Filter capacitor substitution box
   Power resistor substitution box
   Diode rectifiers for substitution testing

Many VOMs provide a scale for DC current measurement; most VTVMs do not. Usually neither has provision for measuring AC current. VOMs are safe for working with "transformerless" power supplies. Because they do not have to be connected to the AC power line during use, there is no hazard of inadvertently shorting across the power line when working on a "hot" chassis. VOMs are portable, self-contained instruments.

A drawback of VOMs is that they are more susceptible to burning up the meter than are VTVMs or FETVOMs if the user forgets to switch to the proper scale or function before connecting the instrument to a voltage-carrying circuit. For this reason, many prefer to use a VTVM in checking power supplies. FETVOMs have some of the desirable characteristics of both VOMs and VTVMs and are often portable. VTVMs or FETVOMs are generally preferred in some uses because their high input impedance has less tendency to load high resistance circuits and produce erroneous voltage readings.

Most isolation transformers do two important things: vary the AC line voltage to equipment under test and eliminate the

1-2 Equipment for Troubleshooting Power Supplies 29

shock hazard of a transformerless chassis. Most power supply problems can be solved without using an isolation transformer, but without it the troubleshooter must be more careful, knowledgeable, and ingenious.

An oscilloscope is not absolutely essential for dealing with ordinary power supply problems. However, for determining the nature and amount of ripple with certainty, the oscilloscope is without peer. If the output filter capacitor of a DC power supply decreases in value, for example, the oscilloscope is the best means available for determining the characteristics of any variation of the B+ line that may result from increased power supply impedance as seen by the load. Some TV problems, especially, are best diagnosed by this technique.

Capacitor checkers, DC milliammeters, tube checkers, and diode checkers are very helpful in speeding up work. They provide data that enables a quick, definite identification of bad parts.

One reservation that should always be kept in mind when using instruments has to do with whether or not the conditions of test duplicate the conditions of use. For example, an ohmmeter check for a short between B+ and B− does not subject the circuit to the same voltage as when the equipment is operating normally. So if the trouble only reveals itself when B+ reaches a high level, the problem will not be diagnosed. Or, a damper tube in a TV receiver is another example: No tube checker applies the high pulse voltages that exist across a tube when it is in use. Tube checkers, like any other instruments, are not always reliable when they do not duplicate the conditions of normal use.

A wattmeter is an instrument that deserves greater popularity in troubleshooting. When used in conjunction with a variable AC source (isolation transformer), the wattmeter enables instant monitoring of power input from the AC power line. If the trouble results in too much power input, early awareness and correction can save many dollars in needlessly burned-out new parts.

Substitution boxes that provide a variety of sizes of clip-in components save much time in testing. Capacitor substitution boxes that contain a choice of sizes of electrolytic capacitors and power resistor substitution boxes that can be inserted in place of a filter resistor or used to provide known loads of various sizes are especially helpful in power supply work. The component substitution technique is an old and good one. If substitution boxes are not available, a supply of individual components can be used instead; it will simply take more time and be less convenient.

Where necessary, regulating circuitry is commonly inserted

between the basic supply and the load. Organization of typical supplies is shown in Figure 1-1.

1-3 Basic Troubleshooting Procedures for Power Supplies

How does one troubleshoot a DC power supply? To answer that question it is helpful to think of the power supply in the simplified form shown in Figure 1-2a. Whatever voltage is behind the DC output terminals will act like a battery as well as a source of ripple in series with a resistor to the load. The resistor represents the total effect of wires, diodes, switches, and transformers that act to limit current. The total is generally termed R_S, source resistance.

The DC voltage between B+ and B− *with no load connected* is the battery voltage (V_S). *With the load connected,* the DC voltage between B+ and B− is the battery voltage V_S, *less* the voltage drop across the source resistance: $V_{out} = V_{no\ load} - I_L R_S$. In common rectifier-filter supplies the output voltage is usually 60% to 80% of the output voltage under no load.

The most frequently occurring power supply troubles have to do with too low values of DC output voltage, too high ripple, or too high source resistance or impedance as seen by the load. This is illustrated in Figure 1-2. Abnormally low DC output voltage

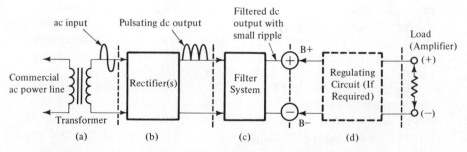

Figure 1-1 Typical power supply organization. (a) Transformer (either on the chassis or the power pole) establishes input voltage level to be rectified. (b) Half-wave or full-wave rectifier changes AC to DC. (c) Filter system stores energy and releases it to greatly reduce ripple. (d) Only where it is required is regulating circuitry added to the basic power supply to further stabilize DC output level to the load.

1-3 Basic Troubleshooting Procedures for Power Supplies 31

and abnormally high ripple often occur together, both resulting from either a defective input filter capacitor or an excessive load.

It is important for the troubleshooter to develop a logical sequence of procedures to save time. Table 1-1 takes a typical problem, low B+ output voltage, and outlines a series of specific steps and measurements to locate the cause of trouble. For convenience and efficiency it is suggested that troubles be divided into two categories: (1) those originating outside the power supply; and (2) those originating inside the power supply.

Of course, the values in Table 1-1 cannot be precisely correct for all power supplies. This is because no one can anticipate the specific values that circuit designers may use in their work. However, Table 1-1 applies to the vast majority of basic power supplies and can be relied upon as a tested, sound procedural guide.

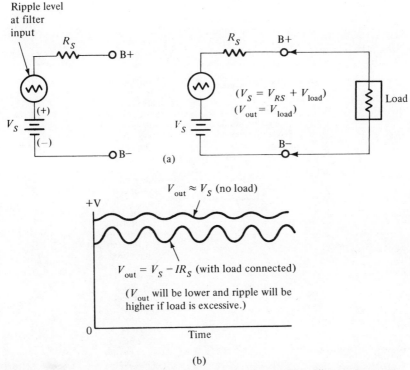

Figure 1-2 (a) Simplified equivalent circuit of a rectifier-filter power supply with no load and with load. (b) Comparison of output voltage under no-load and load conditions.

TABLE 1-1

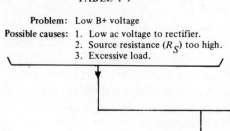

Trouble originates outside power supply

1. *Low ac voltage to rectifier*

   Possible causes:
   a. Low line voltage (occasionally happens).
   b. Power transformer open or shorted, incorrectly connected, or not the right transformer (seldom happens).

   Checks
   With ac voltmeter, measure line and transformer voltages. If incorrect, check to find cause. If correct, proceed to 2.

2. *Excessive load*

   Possible causes:
   ** Short between B+ and B−, improper amplifier bias or faulty adjustment (very common).

   Checks
   1. With ohmmeter, check for short between B+ and B−; R should usually exceed 1000 ohms, often much more.
   2. Measure dc voltage drop across filter choke or resistor and measure R of choke or resistor. If both are normal, load is not excessive. If R is normal but voltage drop is too high, the problem is excessive load.
   3. Disconnect load. Repeat 1 and 2 above to determine if problem is inside power supply or outside power supply. (Load problems will be *discussed separately*.)

1-3 Basic Troubleshooting Procedures for Power Supplies

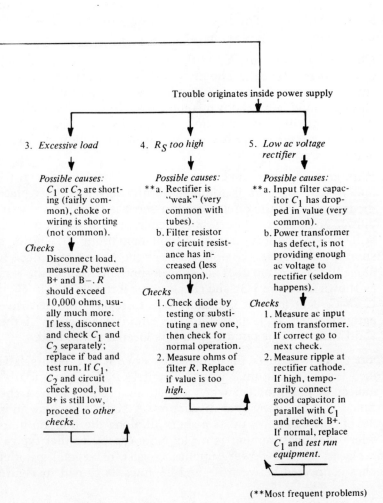

Trouble originates inside power supply

3. *Excessive load*

   Possible causes:
   C_1 or C_2 are shorting (fairly common), choke or wiring is shorting (not common).

   Checks
   Disconnect load, measure R between B+ and B−. R should exceed 10,000 ohms, usually much more. If less, disconnect and check C_1 and C_2 separately; replace if bad and test run. If C_1, C_2 and circuit check good, but B+ is still low, proceed to *other checks*.

4. *R_S too high*

   Possible causes:
   **a. Rectifier is "weak" (very common with tubes).
   b. Filter resistor or circuit resistance has increased (less common).

   Checks
   1. Check diode by testing or substituting a new one, then check for normal operation.
   2. Measure ohms of filter R. Replace if value is too *high*.

5. *Low ac voltage rectifier*

   Possible causes:
   **a. Input filter capacitor C_1 has dropped in value (very common).
   b. Power transformer has defect, is not providing enough ac voltage to rectifier (seldom happens).

   Checks
   1. Measure ac input from transformer. If correct go to next check.
   2. Measure ripple at rectifier cathode. If high, temporarily connect good capacitor in parallel with C_1 and recheck B+. If normal, replace C_1 and *test run equipment*.

(**Most frequent problems)

1-4 Running Down Short Circuits

If the power supply blows fuses repeatedly, smokes or gets excessively hot in a short time, a short circuit in the power supply or the load is indicated. The first thing to do in such a case is to make the following series of checks:

1. Check solid state diodes with an ohmmeter to see if they have low resistance in one direction, high in the other. (Low resistance in both directions indicates a shorted diode.) If a short is found, a breakdown somewhere else may have caused it by permitting excessive diode current. Do not replace diode and turn circuit on until you have made the following check and found no short!

2. With an ohmmeter, check for too low resistance between B+ and B−. If no short is found it is probably safe to replace diodes and turn the circuit on. If a short is found, divide circuit at Point A in Figure 1-3a and check to see which portion still shows a short. If the portion containing C_1 shows a short, disconnect C_1 completely from the circuit and check it alone. If this portion does not show a short, disconnect the load as shown in Figure 1-3b and check to see if the short is in C_2. If C_1 or C_2 is at fault, it is usually safe to replace the defective part and then apply power to the circuit for testing. If the short is not in this circuit, then it is located in the load.

A *short* exists whenever the resistance between two points falls to much lower than normal values. Parallel circuits, particularly, are subject to this problem when something occurs to lower the resistance of one of the branch circuits. A short anywhere in a parallel circuit will show up as a short all along the line, no matter where the measurement is made. Note especially that typical loads placed on DC power supplies consist of several parallel branches connected between B+ and B− lines as shown in Figure 1-4a.

The best procedure to isolate the cause of a short between B+ and B− lines, or another circuit, is as follows:

1. Break (open) the circuit at some convenient point to form two sections, as in Figure 1-4b.
2. Measure the resistance between B+ and B− lines in one sec-

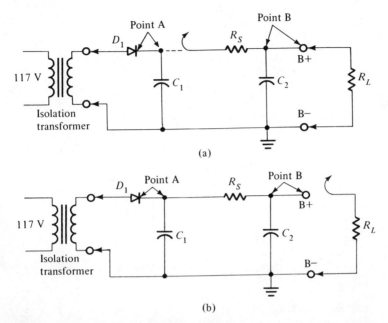

Figure 1-3 (a) To see if C_1 is shorted, disconnect rest of circuit at Point A as indicated, then measure resistance from Point A to ground. (b) If C_1 is good, disconnect the load from Point B as indicated and then measure resistance from Point B to ground to see if C_2 is shorted; if C_2 is also good, the load must have a short in it.

tion and then in the other. One section should show good (high resistance), while the other will still show the short.

3. Break the section of circuit with the short in it at some convenient point. Measure the resistance between B+ and B− lines once more in each new section. One section will show good, while the other will still show the short (see Figure 1-4c).
4. This procedure must be repeated as many times as necessary until the exact location of the short circuit is found.

Note: The above procedure will not usually reveal the cause of excessive load current when it is due to improper bias of the amplifiers (either transistor or vacuum tube) or when circuit adjustments, such as tuning, are set wrong. Solutions to these problems are discussed in later chapters.

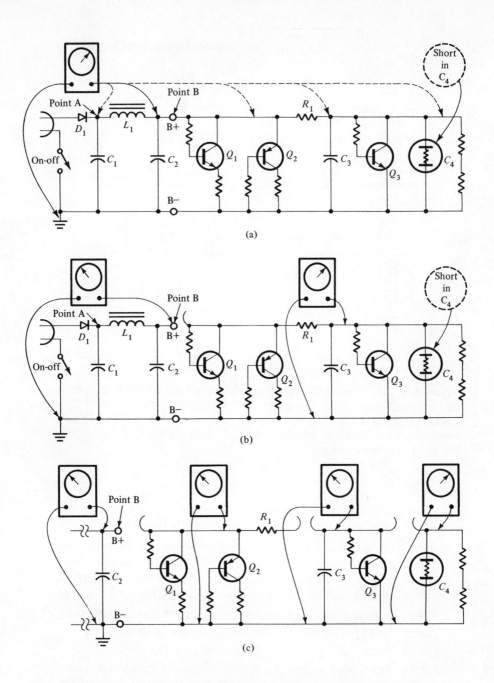

Figure 1-4 (a) DC power supply with simplified load to show parallel arrangement of elements. (Note that a short in C_4 will show up as a short between B+ and B− wherever measured.) (b) Circuit is broken at Point B to determine if short is in power supply or in load. (c) Circuit is opened at other points until trouble is narrowed to the defective part or exact site.

1-5 Half-Wave Rectifier Power Supplies

When loads are small, usually under 10 watts, half-wave rectifiers are popular because they are the simplest and least expensive DC power supply circuits. Small record players, radio receivers, and intercoms are examples of domestic electronic equipment where half-wave circuits are frequently found.

The familiar small table model radio receiver has probably been manufactured in greater numbers than any other piece of home electronic equipment. It normally uses a half-wave rectifier-filter DC power supply typical of other small units. For this reason we have chosen it to examine closely. In vacuum tube receivers a radio normally takes the form shown in Figure 1-5. Because this unit does not have a power transformer on the chassis it is sometimes referred to as a transformerless power supply. But of course the circuit does have a transformer — on the power pole outside the house.

For safety, it is best to use an *isolation transformer* when working on equipment using any type of transformerless arrange-

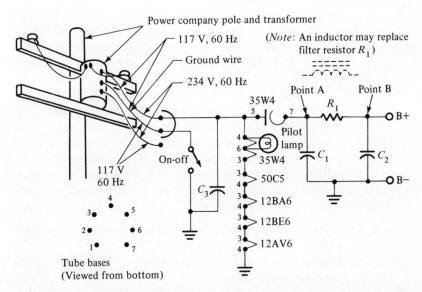

Figure 1-5 Half-wave rectifier DC power supply typically found in tube type table model radio receivers. C_3 is used to bypass any radio signals that may be picked up by the power line acting as a long wire antenna.

ment. Most isolation transformers designed for this use incorporate a means of adjusting the output voltage and a meter to measure it. This is a very useful service feature to have available for troubleshooting many different types of electronic equipment.

Common troubles in the power supply of Figure 1-5 are:

Case 1: A filament, heater, of one of the tubes opens and no tube operates.

Case 2: Rectifier emission decreases (weak tube) causing high source resistance and low B+. The sound is bad: It distorts easily, and may have a 60 cycle hum.

Case 3: The input filter capacitor (C_1) diminishes in value as it ages, causing greater X_C and lower AC input to the rectifier, which causes low B+ and bad sound.

Case 4: The output filter capacitor (C_2) diminishes in value as it ages, causing hum, whistles, and oscillation.

Case 5: A capacitor is leaky or shorted.

Let us see how each case is handled:

Case 1 Trouble: No tubes light up, set is dead; suspect open heater.
- *One solution:* Check tubes in a tube checker, replace dead one, and check for normal operation.
- *A second solution:* Remove tubes one at a time, with an ohmmeter check each for an open heater, replace tube that has open heater, and check for normal operation. (This operation requires only an ohmmeter, and may be the fastest method.)
- *A third solution:* Turn on the radio power switch, and with an ohmmeter, check point to point for a complete circuit, as shown in Figure 1-6. When the open circuit is found, repair and check. (When the circuit is normal, there is a complete circuit from one side of the power plug to the other.)

Note: The third solution will reveal *any* open in the circuit, whether it be line cord, switch, plug or tube heater; the other solutions will detect tube faults only.

Case 2 Trouble: Rectifier weak, low emission, resulting in low B+; sound is bad.
- *One solution:* Check rectifier in a tube checker, replace as indicated, and test run.
- *A second solution:* Remove suspected rectifier, replace with a good one, and check receiver operation.

1-5 Half-Wave Rectifier Power Supplies 39

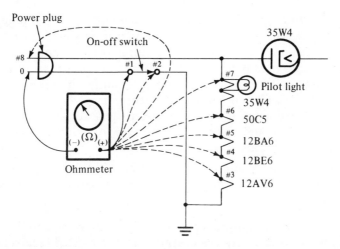

Figure 1-6 Sketch showing sequence of resistance checks to locate an open in the heater circuit of a 5-tube radio receiver. If the circuit is good, the ohmmeter should show a complete circuit from one side of the power plug to the other when the power switch is on, points 0-#8.

Case 3 Trouble: Input filter capacitor C_1 has dried out; B+ is low as a result, and sound is bad.
- *One solution:* Connect a good capacitor in parallel with the suspected one, turn set on, and check B+. If B+ comes back up to normal, replace C_1 and test run receiver.
- *A second solution:* Check amount of ripple at Point A (Figure 1-5); if above normal, replace C_1 and test run receiver.
- *A third solution:* Check characteristics of C_1 with a capacitor checker, and replace as indicated. (C_1 must be disconnected from the circuit to eliminate possible parallel paths which may disturb readings.)

Case 4 Trouble: Receiver hums, whistles, or oscillates due to a dried-out output filter capacitor C_2.
- *One solution:* Turn set on and the volume up until trouble shows; then connect a good capacitor in parallel with C_2 and note if trouble symptoms disappear. If set operates normally, replace C_2 and test run the receiver.
- *A second solution:* Turn set on, and the volume to the lowest point. Measure B+, the DC output voltage of the power supply; this should be near normal. Connect the oscilloscope to Point B (Figure 1-5) and observe ripple. Slowly advance vol-

ume while watching oscilloscope carefully to see if audio signal (program material) begins to appear as sound gets louder. Connect a good capacitor in parallel with C_2 and note effect on ripple as shown on the oscilloscope. If ripple returns to normal (without audio showing, even with volume turned up), replace C_2 and test run the receiver.
- *A third solution:* Check C_2 with a capacitor checker, replace as indicated, and test run the receiver. As with C_1, disconnect C_2 for checking.

Case 5 Trouble: A capacitor is leaky or shorted.
- *Solution:* Disconnect and check with ohmmeter or by substitution.

The same general procedures outlined above and in Table 1-1 also apply to transistorized receivers. Transistorized receivers usually require lower values of B+ voltage. Where these receivers use half-wave rectifier power supplies, there is usually a step-down transformer on the radio chassis to provide the lower DC voltage required. Too, transistors do not require heating, so there will be no heater circuit.

Use of the foregoing solutions and those in Table 1-1 will enable almost anyone to solve nearly all straightforward half-wave power supply problems.

1-6 Full-Wave Rectifiers Using a Center-Tapped Transformer

Before the advent of semiconductors, which make *low power* bridge rectifier circuits feasible, the full-wave rectifier using a center-tapped transformer was popular for medium power applications. The circuit requires two diodes, as shown in Figure 1-7. Usually both diodes are constructed in a single unit known as a duo-diode. This circuit is both more efficient and easier to filter than half-wave circuits.

A comparison of the outputs of half-wave and full-wave rectifiers when connected to a resistive load *without filtering* is shown in Figure 1-8. There are other common rectifier circuits, such as the half-wave voltage doubler (transformerless), the full-wave voltage doubler (transformerless), and the full-wave bridge rectifier (with a transformer). (Note that all circuits are classified either as

1-6 Full-Wave Rectifiers Using a Center-Tapped Transformer 41

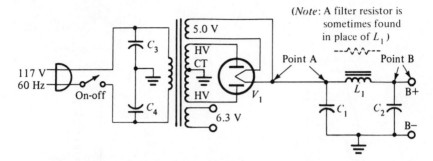

Figure 1-7 Standard full-wave rectifier circuit using duo-diode rectifier tube and center-tapped transformer. Transformer includes a 6.3 V winding to heat tubes of amplifier (load) which will be connected in parallel across it.

half wave or full wave depending on whether one or two output pulses per cycle are fed to the filter system.)

Common troubles in full-wave rectifier power supplies:

1. High source resistance causing low B+ output due to a weak rectifier tube or an open solid state diode.
2. Low B+ output caused by a decreased value of input filter capacitor (C_1).
3. Equipment that oscillates, distorts, and/or operates erratically due to a decreased value output filter capacitor (C_2) which creates too high power supply impedance as seen by the load.
4. Less frequently, troubles resulting from a defective power transformer or filter choke that is shorting or open.

We note that these troubles are the same as those discussed earlier for half-wave circuits. *Solutions to full-wave rectifier supply problems require the same procedures outlined in Table 1-1 and Section 1-5.*

Once again, we call attention to the fact that problems in the load which cause it to draw excessive current from the DC power supply have the same result and symptoms for all basic types of DC power supplies. The immediate problem that faces the troubleshooter when he finds low B+ is how to determine as rapidly as possible whether the trouble lies within the power supply or in the load. If procedures of Table 1-1 are followed the location of

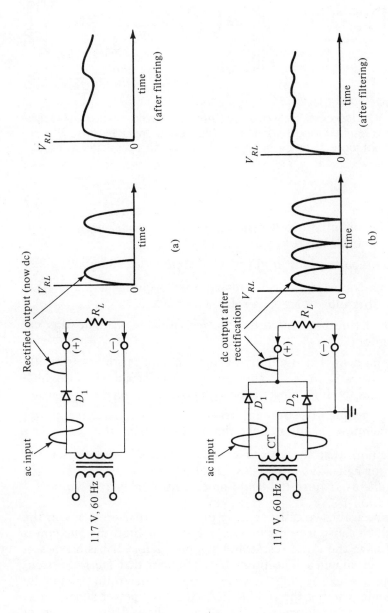

Figure 1-8 Comparison of output of (a) half-wave and (b) full-wave rectifiers without any filtering, or if filtering should fail completely. Effects of filtering are shown at extreme right.

the trouble can normally be found after a short sequence of checks.

1-7 Bridge-Rectifier Power Supplies

The bridge rectifier is a full-wave rectifier which is more efficient than half- or full-wave circuits using a center-tapped transformer. In fact, it is the most efficient basic rectifier circuit in common use. As shown in Figure 1-9, this circuit delivers full-wave rectified output to Point A, the filter input, in the same way as the full-wave circuit using center-tapped transformer shown in Figure 1-8. The filter is the same.

The fact that the bridge circuit requires four diodes is a disadvantage when vacuum tubes are used, because of the additional filament wiring and space required. Hence, vacuum tube bridges are seldom used in small power supplies. The vacuum tube rectifier has long dominated high power applications because its higher efficiency outweighed its disadvantages. Solid state diodes do not have the shortcomings of vacuum tube diodes. Solid state bridges are commonly available today in a single unit package. Small and inexpensive, they are now being used in many medium- and low-power applications.

How does the bridge circuit work? Diodes on opposite legs of the transformer conduct in series at the same time. One pair of the diodes conducts on one-half of the input AC cycle and the other pair conducts on the next half cycle. In this way, full-wave rectification is accomplished.

Special problems with bridge rectifiers arise when one of the diodes shorts. This places one of the diodes from the opposite pair

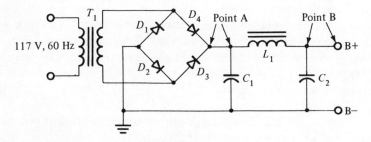

Figure 1-9 Bridge-rectifier circuit with pi type filter.

directly across the power transformer secondary winding, with almost nothing in series with the good diode to limit current when it is forward biased (in conduction). Needless to say, something has to break down or burn up when this occurs: the fuse, the other diode, the transformer, or the power switch.

To test the rectifiers in a bridge, it is necessary to disconnect the circuit wherever it constitutes a parallel path. When it is certain that the diodes have been isolated from anything in parallel that may disturb the reading, a correctly functioning diode should show very high resistance in one direction and much lower resistance in the other direction (when the ohmmeter leads are reversed).

When diodes are found to be defective, power should not be turned on after defective units have been replaced until one has checked for a short between B+ and B− lines. Even if no short seems to exist, a good precaution is to first apply *low* voltage from an isolation transformer, then advance the voltage slowly, watching carefully for any sign of too rapid heat build-up anywhere in the equipment.

1-8 Half- and Full-Wave Voltage Doublers

Voltage doublers evolved as a method of producing approximately 240-260 volts DC under load from a standard 117 volt AC power line without using a transformer on the chassis. As mentioned previously, such power supplies are often referred to as "transformerless," although actually the power transformer is still in the circuit, but outside the house on the power company's pole instead of on the chassis.

The popularity that doubler type power supplies enjoyed for vacuum tube circuits seems unlikely to carry over for transistor circuits. That is because vacuum tubes work best at voltages over 100 V while common transistors work best at voltages below 30 V. We discuss doubler power supplies here only because there are still many TV receivers and audio amplifiers in use which employ vacuum tubes and use doublers.

Caution: All transformerless power supplies present potential hazards to the unwary troubleshooter and to his equipment. If the power plug of the transformerless unit is inserted "upside down" into the wall socket, the chassis is "hot" rather than at ground potential, as shown in Figure 1-10. If the ground lead of a piece of test equipment is connected to the chassis (as is common practice), and the chassis is "hot" because of the reversed power plug, *the ground lead that has been connected is actually a short circuit directly across the AC power line!* It is both startling and frightening to see a flash as a lead vaporizes right before our eyes. To remove this shock hazard, it is recommended that an isolation transformer be used when working with such power supplies.

Full-wave voltage doublers are more popular than half-wave doublers. Circuits of both, that have been used in TV receivers, including heater circuits, are presented in Figure 1-11.

Typical troubles in voltage doubler power supplies are:

1. Input filter (doubling) capacitors, such as C_1 and C_2 of Figure 1-11, decrease in value as they age, which lowers B+ and increases ripple. A shrunken TV picture with a dark border is a typical result.

2. Diodes often short, open, or develop higher than normal resistance. This is frequently the result of excessive current, caused by a short or circuit fault somewhere else. Check for shorts, and be alert for overloads when testing after replacing defective diodes.

3. Output filter capacitor C_3 decreases in value, raising power supply impedance, causing high ripple, oscillation, and erratic operation.

The same methods and techniques of troubleshooting that were discussed for the transformerless half-wave rectifier circuit (Section 1-5) should be used here. If *one* of the doubling capacitors (C_1 or C_2) in a full-wave circuit is found defective, it is good practice to replace *both* C_1 and C_2. This avoids an imbalance in capacitor size which subjects the smaller, dried-out capacitor to overvoltage, inviting another breakdown in the near future.

1-9 Voltage Regulation

Some types of loads require that B+ be held quite steady, with little variation permitted. Other types of loads tolerate poor voltage regulation with few noticeable ill effects on equipment performance.

Most home electronic equipment does not require a regulated power supply. A simple, basic DC supply such as one of those dis-

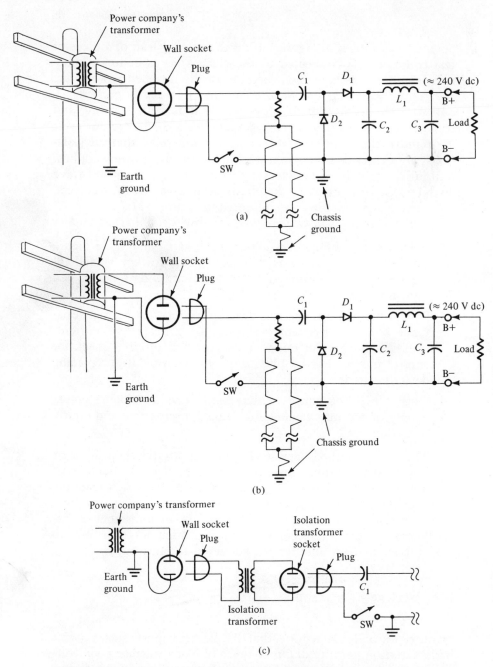

Figure 1-10 Half-wave voltage doubler circuits (a) Power plug "right side up" and chassis at true ground potential. (b) Power plug "upside down" and chassis hot. (c) Recommended way to avoid hot chassis by using isolation transformer when working on any transformerless type of power supply and its load.

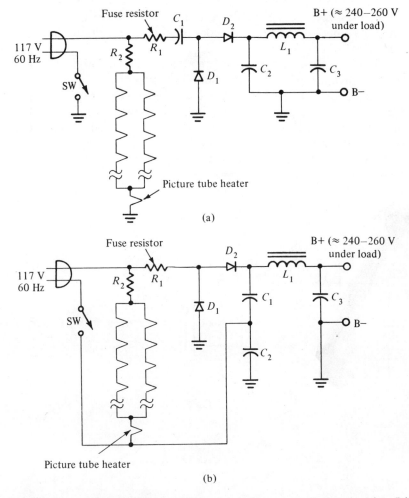

Figure 1-11 Voltage doubler circuits. (a) Half-wave voltage doubler. (b) Full-wave voltage doubler.

cussed earlier in this chapter is satisfactory and that is what is generally used. In most small radio receivers, TV, tuners and some audio frequency (AF) amplifiers the amount of DC current called for from the power supply is nearly constant. Put another way, these units have single-ended power amplifier stages which operate Class A and therefore represent a constant DC load on the power supply.

If we refer to the equivalent circuit of Figure 1-2, we see that a constant load produces a constant DC voltage drop across the source resistance (R_s). B+ voltage must then remain constant and steady, as long as V_s does not change due to power line fluctuations or other problems.

Changing loads are another story. They do not produce a constant voltage drop across the source resistance, and hence B+ changes as the load changes. It is here that some form of voltage regulation may be required. Push-pull AF power amplifiers operating at or near Class B are the usual cause. (Class B push-pull is used at high power to increase efficiency and lower distortion.)

Regulating techniques may be applied in several ways. Two of the most popular are sketched in Figure 1-12. For some small loads in which all stages are operated at Class A except the Class

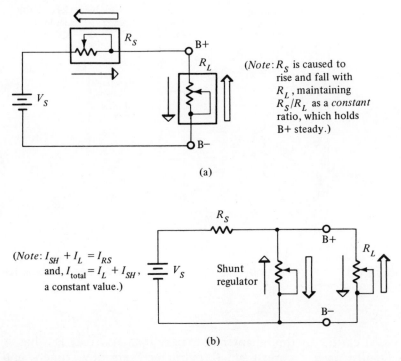

Figure 1-12 (a) Series regulator circuit in which $V_L/V_S = R_L/R_{total}$. As long as the resistors hold a constant ratio the voltages must hold the same ratio and V_L will be constant as long as V_S does not change. (b) Shunt regulator circuit in which shunt current is caused to rise as load current falls and fall as load current rises. Together, $I_{SH} + I_L$ equal a constant load current through R_S and hence a constant voltage drop across R_S with the result that V_L will be constant as long as V_S does not change.

B push-pull final AF amplifier, regulation is applied to the stages preceding the final amplifier but not to the final itself. Pentode vacuum tubes and transistors which have similarly shaped characteristic curves can tolerate moderate changes in plate and collector voltages, especially in push-pull, without those changes significantly affecting the output signal. Shunt (parallel) regulation is usually employed in this application, as shown in Figure 1-13. A Zener diode is most often used when the load is made up of transistor circuits, and a voltage regulator (VR) tube is used when the load is made up of vacuum tube circuits. In either case, an amplifying transistor or a vacuum tube may be used in place of the Zener diode or the VR tube.

The load on the basic power supply in shunt regulation con-

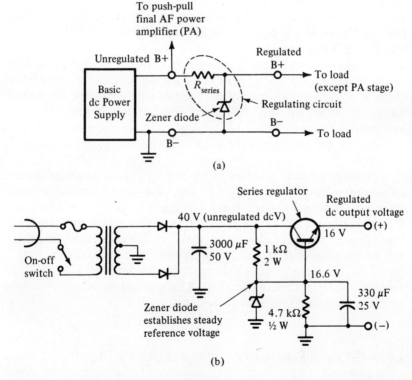

Figure 1-13 (a) Basic shunt regulator circuit using a Zener diode. In higher voltage circuits the Zener diode is often replaced by a VR (voltage regulator) tube. (b) Basic series regulator circuit using a series "pass" transistor as the variable resistance element which changes in value as the load changes.

sists of two elements, Zener (or VR) current and load current; thus, $I_{\text{basic } PS} = I_Z + I_L$. If the Zener or VR tube becomes open, regulation is lost, load on the basic supply is less, and the B+ voltage tends to rise because the IR_S drop across the source resistance becomes less. If the measured value of regulated B+ is found to be higher than specified, the Zener or VR tube should be checked, as well as the series resistor through which $I_Z + I_L$ flows.

A short in the regulating circuit is a short in the load on the basic power supply, and appropriate procedures should be instituted for running down a short. (See Table 1-1 and Section 1-4 for suggestions.)

Series regulation uses an amplifying device, either a transistor or a vacuum tube, as the series element whose resistance is controlled to form a constant ratio between R_S and R_L. This particular series element is a power amplifier transistor or vacuum tube operating as an emitter follower or cathode follower and the load on the power supply is the emitter resistor or cathode resistor, as the case may be. Sometimes the series amplifier is preceded by additional amplifiers which drive it if the regulation must be especially good. Series regulation circuits normally include a reference voltage against which the regulated B+ output voltage is compared; any difference (or variation in the difference) becomes the signal which ultimately controls the resistance represented by the series transistor to hold the R_S/R_L ratio constant as R_L varies.

Troubleshooting a series regulation circuit *is* troubleshooting an amplifier circuit. The signal in this case is any variation in the difference between the reference voltage (a steady DC voltage) and B+, no matter whether the variation is of short or long term duration. One of the first checks to make is the reference voltage itself to be sure that it has not changed for some reason. If it is found to be correct, standard procedures for checking an audio amplifier should then be applied. These procedures are discussed in detail in later chapters.

1-10 Ground Points and How to Locate Them

Anyone is at a great disadvantage in dealing with electrical and electronic circuits if he does not understand clearly what is meant by ground, and how DC and AC ground points may differ.

1-10 Ground Points and How to Locate Them 51

Like zero in our number system, ground in an electrical or electronic system represents a zero point, a point of reference against which all other points of the system can be compared. By custom, we apply this way of looking at things to voltage in electricity and electronics. *Thus, ground is a point of zero voltage in a circuit;* voltages at other points are positive or negative with respect to ground. Schematic drawings of electronic equipment usually include DC voltage values at transistor and tube terminals as well as at other strategic points. *Such DC voltage values are always with respect to ground, unless stated otherwise.*

To measure DC voltages in an electronics system, the ground lead (sometimes called *negative,* or *common*) of the DC voltmeter should be connected to a ground point of the circuit. Note that all ground points drawn in a schematic are actually all connected together and form one point: no matter which ground point a troubleshooter connects to, it will be the same as the others, and valid measurements can be made.

Naturally, a troubleshooter should seek the most accessible and convenient ground point when connecting an instrument. Some of the best "landmarks" to find ground are filter capacitors, volume controls, and power transformers. These are large, distinctive, and easily identified components, and each will usually have at least one connection to ground. (Refer to the schematic of the equipment to verify.) Figure 1-14 shows how to locate ground in most home electronics units.

In practice ground is treated differently in large AC power distribution systems from many electronics systems. In most large AC power distribution systems the ground point is just that, a connection to the earth itself. This is accomplished by connecting to underground metallic pipes of a water system, a stake driven into the ground, or some other method that will insure a good connection to the earth. In portable power systems and in electronics systems the ground point may not be "earth" ground but merely a common reference point for that particular system. *That is, the equipment ground and earth ground may not be at the same voltage because they are not connected to each other.*

All of our precautions in dealing with transformerless supplies stem from the possibility of equipment ground and earth ground being at different voltages. This depends upon the circuit, and/or whether the power plug is inserted "right side up" or "upside down." Use of an isolation transformer greatly reduces the possibility of inadvertently connecting two different grounds to-

52 Troubleshooting DC Power Supplies

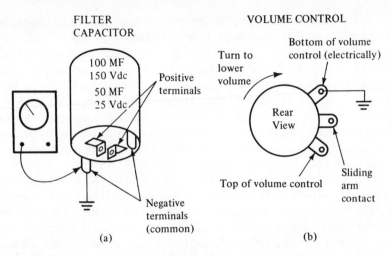

Figure 1-14 Easily found components that aid in locating ground points. (a) Filter capacitor. (b) Volume control.

gether, each having a different voltage, and thereby producing a damaging short. The point to remember is that a piece of electronic equipment has its own ground point and great care must be taken when connecting instruments to make sure that the instrument panel is not at a different voltage! One piece of equipment cannot tolerate two different ground potentials.

AC ground points do not always coincide with DC ground points. This is extremely important to keep in mind. The designated ground point in a particular piece of equipment is always a ground point for both AC and DC voltages. There are, however, additional points that will be at AC ground potential but not at DC ground potential. For example, one of the output terminals of a DC power supply is normally connected to ground while the other is at a positive or negative DC voltage (B+ or B−) with respect to that ground. But both output terminals of the DC power supply are AC ground points as long as the filtering is effective and the terminals do not vary significantly in voltage with respect to each other. (The small amount of ripple is normal and is considered insignificant.)

By-pass capacitors, are intended to establish the connection points at both ends of the capacitor at the same AC level so that one end does not vary significantly in voltage with respect to the other end, even though the DC voltage levels at the two ends are usually quite different.

AC waveforms may be shown in schematics as variations above and below a zero axis with no reference given as to whether the axis itself is at actual ground potential or at some DC level other than ground. Or they may be shown as variations with respect to ground. This is illustrated in Figure 1-15. In nearly all electronic equipment, it is standard practice for the signal to progress through the equipment as variations of pre-established DC levels we have referred to previously as essential static conditions. These must first be proper before desired amplification can be achieved.

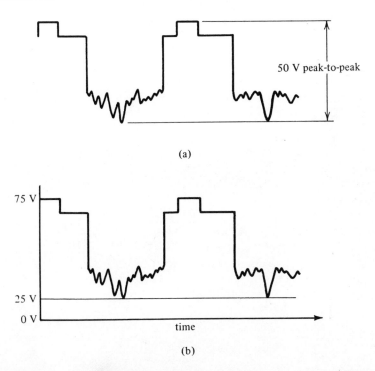

Figure 1-15 (a) AC waveforms in equipment may be shown with only peak-to-peak values. (b) The same waveform may be shown as variations of the DC levels present in an amplifier.

chapter 2

Audio amplifiers in radios, television receivers, and other equipment

The audio frequency (AF) amplifier forms one functional section of most home electronic equipment. As its name implies, this part of the total system amplifies signals we are capable of hearing, in the approximate frequency range from 14 to 20,000 Hz. The AF amplifier is an integral part of radio receivers, television receivers, high fidelity systems of all types, and intercommunication systems; in fact, it is an essential part of any system designed to amplify voice or music. Although we seldom think of it because we do not see the amplifiers, our telephone system depends heavily upon audio amplifiers to overcome losses in the lines, switches, and other parts of the system.

Basically, the AF amplifier is expected to fulfill the following functions:

1. *To accept the signal at whatever input voltage and current level is available from transducers and signal sources in accordance with its design specifications.* Signal input levels commonly range from a few millivolts to approximately one volt. There are exceptions, of course, which can be determined by referring to equipment specifications.
2. *To amplify the signal to the desired power level with minimum distortion.*
3. *To perform its function in a stable, prompt, and dependable manner with minimum attention and maintenance over a long period of time.*

Trouble, then, in an AF amplifier means that it is failing to fulfill its expected function in one or more of the following ways:

56   Audio Amplifiers in Radios, Television Receivers

1. By distorting the signal in some way.
2. By failing to produce its specified power output when the input signal is normal for that particular piece of equipment.
3. By operating intermittently or erratically under normal environmental conditions (temperature, vibration, dust, humidity, air pressure, etc.)

With these observations in mind, we shall now look at details of analyzing and repairing some typical AF amplifiers.

2-1   Initial Considerations in Troubleshooting AF Amplifiers

In radio and television receivers the AF amplifier is only a part of the total system. One of the first steps in troubleshooting amplifiers (or any malfunctioning system) is to "localize" the trouble by narrowing it down to one section, functional block, or stage as quickly as possible.

To begin, we consider the functional organization of the system. Figure 2-1 presents the block diagram of a typical small radio receiver, descriptive of either tube or transistorized type. Note that the information-carrying signal enters the antenna as RF (radio frequency) energy, then is converted to IF (intermediate frequency), which is a lower frequency than the RF in the antenna, but higher than the AF which follows detection. The IF circuits, once set by proper alignment (adjustment) procedures, remain fixed in frequency as different stations are tuned. Output of the IF amplifier is fed to the diode detector. This recovers the original audio frequency signal and delivers it as a varying signal (AF) voltage across the volume control.

The sliding contact of the volume control enables us to choose the amount of signal voltage we send on to the AF amplifier. The amplifier-output signal level is customarily controlled by this controlling of the input signal level. The amplifier gain itself is usually fixed by the design and does not change. (Remember gain is a *ratio* of output to input.)

The first check of a typical tube-type radio that is dead is to see if the tube heaters light up. In the series circuit (shown in Figure 2-2), none of the tubes will heat its cathode and operate if there is a break anywhere in the circuit. To locate the open circuit, refer to procedures outlined earlier in Section 1-5. There is

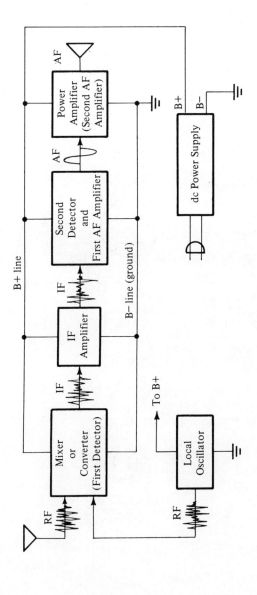

Figure 2-1 Functional organization and block diagram of a small radio receiver. Additional stages are sometimes added for more gain.

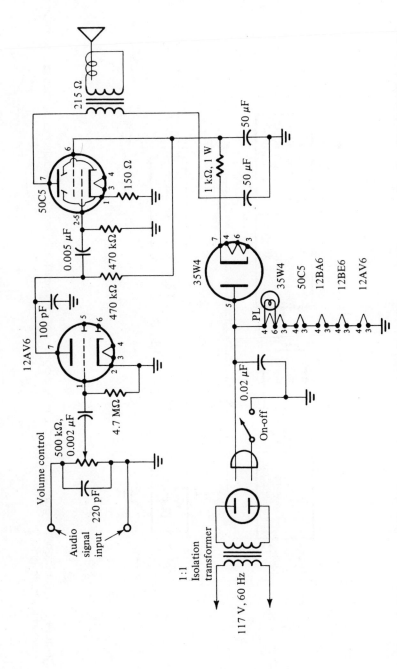

Figure 2-2 Audio amplifier and power supply of typical small tube type radio receiver. Note that an isolation transformer is recommended when working on any transformerless piece of equipment.

2-1 Initial Considerations in Troubleshooting AF Amplifiers 59

little to add to these procedures except a hint to save time. As in any kind of electronics troubleshooting, *do not physically remove the chassis or dismantle the unit until it is clearly unavoidable.* In this case the chassis need not be removed initially; instead, each tube can be removed separately, its heater checked with an ohmmeter and the tube replaced if it shows continuity. If all tubes indicate continuity, the next check is from the power plug to pin 5 of the 35W4 tube socket (with the tube removed, of course). If this shows continuity, check from the other power plug terminal to ground (pin 3 or 4 of the 12 AV6 socket, with the tube removed), with the switch "on". All these checks *should* show continuity; if they do not, you will have to "pull" the chassis to reach the switch terminals. With the chassis removed, the switch can be checked for proper opening and closing. Now you also have access to *both* ends of the line cord so it can be checked for continuity. Line cords frequently open at the molded plug end due to repeated bending at that point. When an open line cord is found, cutting off approximately 2 inches and replacing the plug will usually correct the problem.

If the heater circuit is satisfactory and the tubes light up, but the set is dead, turn the power on and let the tubes warm up for a minute or two. Then, advance the volume slowly to maximum as you listen very carefully for an increase in background noise and static. If heard, the trouble may be in the section preceding the AF amplifier, perhaps from an inoperative oscillator. The presence of noise suggests that the AF amplifier and power supply must be at least partially operative.

At this point, you may check the tubes in a tube checker or by substitution (if noise increased in the previous check). Or you may pull the chassis to check the B+ and the power supply. (A troubleshooter always feels somewhat foolish, though, if he has pulled the chassis and subsequently finds that the problem is only a defective tube.)

If all tubes and visible connections have been checked and are good, but the set is dead, you now *must* pull the chassis to provide access to different points within the circuit for further checks and measurements. When the chassis is out of the cabinet, the following sequence of checks can be instituted, if you have not already done them:

1. As a quick check turn volume to maximum, introduce a 60 Hz hum signal at the top of the volume control as shown in Figure 2-3. A fairly loud 60 cycle hum from the speaker tells

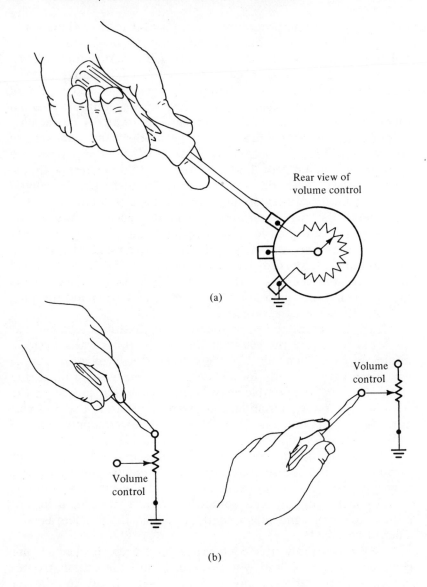

Figure 2-3 (a) Introducing 60 Hz hum signal, picked up by your body from the AC fields in the air produced by power lines, to the top of the volume control as a quick check on AF amplifier operation. (b) Same idea illustrated by using schematic symbol for volume control rather than physical sketch.

2-1 Initial Considerations in Troubleshooting AF Amplifiers   61

the troubleshooter that *both* the AF amplifier and the power supply are operating, probably satisfactorily, and the trouble is probably in the RF or IF section. If no hum is heard, or if it is very faint, the trouble is in the AF amplifier or the DC power supply.

2. With a DC voltmeter, check the power supply for normal B+ output. If B+ is more than 10% below normal, check for a larger than normal voltage drop across the filter resistor, which would indicate excessive load. (The objective here is to determine if low B+ is the result of trouble inside the power supply or in the load. Refer to Table 1-1, if necessary, for steps to follow.)

If B+ is normal, or if the trouble causing low B+ appears to be in the load, institute the series of checks suggested in Table 2-1. This table is a flow chart suggesting a series of checks and measurements to isolate the cause of troubles in the AF amplifier. The exact sequence may be changed at your option if you feel it will be more convenient or will save time. The important thing is to follow some kind of orderly procedure that includes all the necessary measurements and checks.

The preferred method of making DC voltage checks is to connect the negative or common lead of the DC voltmeter to ground of the chassis where it will remain while voltages at different points are being checked. Remember that *DC voltages stated on schematic drawings are with reference to ground unless indicated otherwise.* (For a fuller discussion of ground, refer to section 1-10.)

When a voltage measurement across a particular resistor or other component is needed, one recommended practice is to measure the voltage at both ends with respect to ground, then subtract. The difference is the voltage across the component.

When the final AF amplifier preceding the speaker consists of a single tube or transistor, it must operate Class A. This means that plate or collector voltage and current under static, no signal conditions must be high enough to increase and decrease symmetrically (equally) without reaching cut-off or entering saturation. Furthermore, the voltage and current swings up and down from the static points must represent the AC signal power output that is desired of the amplifier.

The vast majority of single-ended, Class A amplifiers operate in the common cathode or common emitter (CE) configuration,

TABLE 2-1

Trouble: Audio amplifier is dead, low in volume, or noticeably distorts the signal.

Possible causes:
1. Trouble in power supply.
2. A transistor or tube is defective.
3. A capacitor is leaky or shorting or perhaps changed in value.
4. A resistor has changed in value.
5. A connection is bad or intermittent.
6. The signal source or output transducer is not working properly.

General Procedure:
a. Perform dynamic tests to localize trouble to the defective stage.
b. Perform static tests: measure dc voltages, component values, and connections in suspected area revealed by dynamic tests.
c. Analyze data obtained by measurements, correct any problem revealed by data, then test run the equipment to confirm the repair.

a. *Perform dynamic tests*
1. Use signal generator, signal tracer or oscilloscope to see where the trouble first shows up. (Refer to Section 2-2)
2. When trouble stage is found, go to static tests (b) to pinpoint cause.

b. *Perform static tests*
1. Measure dc voltage of power supply and at transistor or tube terminals.
2. Compare measured values with *what should be there*. If not within ± 15%, check further for cause. Substitute new part for suspected part and recheck measurements.
3. When measurements seem to be normal and the unit operates, go to final repair procedure (c).

c. *Analyze data, make repair, and test*
1. Did the defect found occur by itself, or did a different fault cause it? Check possibilities.
2. If no further static discrepancies are found, make the part replacement permanent.
3. Turn on equipment, watching carefully for signs of overheating or other problems. If problems are still present, return to further static checks.
4. If no further problems appear, heat run the unit for several hours to be certain that operation is now normal.

INSTRUMENTS:
VOM, VTVM, or FETVOM (required).
AF signal generator and/or AF signal tracer and/or oscilloscope (necessary if dynamic testing is to be performed).
Transistor checker, tube checker, substitution boxes and/or parts (helpful to speed work).

INFORMATION:
Schematic drawing required at the minimum, with voltages and parts values given. Complete specifications and service notes are best if available.

the best choices when a single tube or transistor is used. Most of such power amplifier circuits use an output transformer with one end of the primary winding connected to B+ and the other end connected to the plate of the vacuum tube or the collector of the transistor. The transformer serves to match the output stage impedance to the speaker impedance for maximum transfer of power.

2-2 Signal Tracing and Signal Injection: Dynamic Checking of AF Amplifiers

Even though final pinpointing of the exact cause of trouble in an electronic system almost always involves checks of *static* conditions, much time can be saved in many cases by using a combination of static and dynamic tests.

There are two basic approaches to troubleshooting using dynamic methods: signal tracing and signal injection. Each places the amplifier under actual signal conditions. It is common practice to use both methods together if instruments are available and convenient. In signal tracing, an oscilloscope or some type of signal tracer is used to examine the signal at each point as it progresses through the amplifier, beginning at the signal input terminals and ending at the speaker. In signal injection, a signal source (usually a signal generator) provides a signal whose frequency and amplitude is adjusted by the troubleshooter. This is injected into the faulty amplifier, beginning at the final stage signal grid or base terminal.

The advantage of signal injection is that the signal generator provides a definite, controllable signal of regular characteristics. Most AF signal generators provide sine wave output and many also provide square waves. In contrast, the signal that is present when normal voice or music is amplified contains a complex, irregular mixture of frequencies, which is more difficult to evaluate than a controlled signal.

In signal injection, test frequencies of 400 or 1000 Hz are commonly used, employing a sine wave or a square wave at the tester's option. When an oscilloscope is used as the signal tracer, a visual assessment of distortion and nonlinearity can be quickly made. Both methods of troubleshooting are illustrated in Figure 2-4.

Note that in signal injection the AF signal is first fed into the grid or base of the *final* amplifer tube or transistor to test the last

64   Audio Amplifiers in Radios, Television Receivers

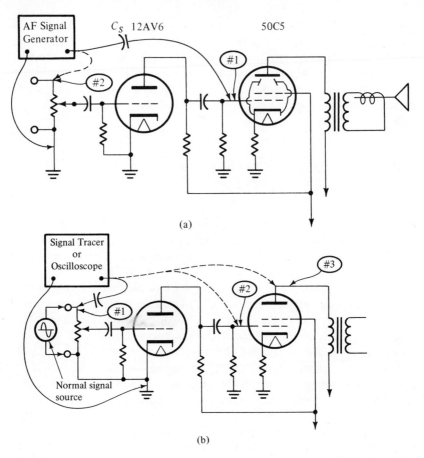

Figure 2-4 (a) Dynamic testing by the signal injection method showing sequence beginning at the output and proceeding step by step to the input. (b) Dynamic testing by the signal tracing method starting at the input (using the normal signal source) and proceeding step by step to the output. Note series capacitor C_s placed in "hot" lead of generator to avoid any possibility of upsetting DC conditions in either instrument or amplifier.

stage operation first. If the sound that is heard indicates that the final amplifier is working properly, the signal generator lead carrying the AF is then moved to the grid or base of the preceding stage. Here, the sound should be louder, and the injected signal can be reduced in voltage. This process is repeated until the defective stage is revealed.

In signal tracing, an oscilloscope or a small, high gain AF amplifier (called a signal tracer) incorporating a volume control and speaker is used to check the progress of the signal through the amplifier *from its beginning at the signal input point.* That is, the amplifier is turned on and adjusted as for normal use and the signal tracer is connected across the signal input terminals to verify that the signal is indeed there. If it is, and if it appears satisfactory, the signal tracer lead is next moved to the grid or base of the first AF amplifier to verify the presence of the signal at this point. If it is there and appears satisfactory, the tracer lead is moved to the grid or base of the next amplifier where the signal should be louder. In this way both the presence of the signal and the distortion can be checked all the way from the input to the output. This procedure also helps to identify defective signal sources and output transducers.

2-3 Using Static (DC) Voltage Measurements as Trouble Clues

Up to this point we have suggested sequences of dynamic checks and measurements to localize the trouble stage by signal injection and signal tracing methods. We have also suggested a sequence of static (DC only) checks in Table 2-1 aimed at further pinpointing the trouble. But remember, our basic thesis has been that proper signal amplification is impossible to achieve unless the proper static conditions exist first.

It is time now to discuss what static measurements tell us about typical small AF amplifiers. Figures 2-5, 2-6, and 2-7 are three examples, one drawn from the table model radio we have been discussing, one from a tube type TV receiver and one from a transistorized car radio. Each diagram has comparable points labeled which are referred to in the following pages. The discussion itself focuses on what clues can tell us when voltages are found to be higher or lower than normal. (We assume that B+ is normal or, if not, that the trouble causing low B+ is in the load.)

Point 1. Check to be sure there is DC output of power supply. If the output is higher than normal, the output AF amplifier may be drawing *too little* I_P or I_C. If it is lower than normal the output AF amplifier may be drawing *too much* I_P or I_C. Check further.

Point 2. Check DC voltage at plate or collector, subtract from reading at Point 1a in each diagram. If the difference is *higher*

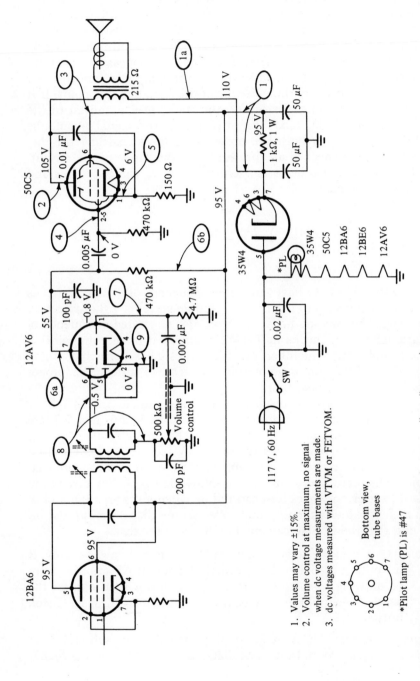

Figure 2-5 Audio section of tube type small radio receiver.

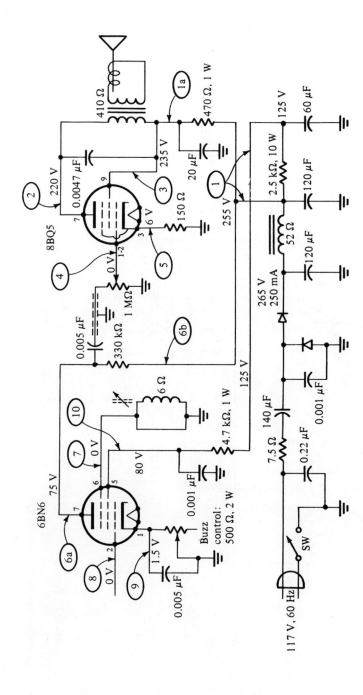

Figure 2-6 Audio section of tube type television receiver. Note half-wave voltage doubler type DC power supply.

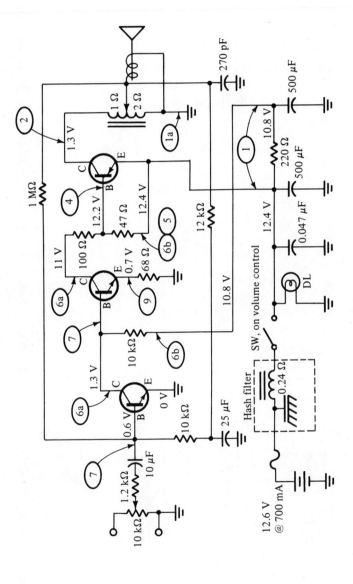

Figure 2-7 Audio section of transistorized auto radio. Note that numbers 3, 8, and 10 are omitted since transistors have no comparable points to these points on vacuum tubes in Figures 2-5 and 2-6. Also, points 6a, 6b, and 7 on this drawing are repeated because the points are comparable.

2-3 Using Static (DC) Voltage Measurements as Trouble Clues

than normal, *too much* I_P or I_C is indicated. If the difference is *lower* than normal, *too little* I_P or I_C or a short across the winding is indicated.

Point 3. This check applies only to vacuum tubes that have a screen grid. If the voltage on the screen grid is *higher* than normal, it indicates that the DC voltage from the power supply is *too high* or the IR drop across the screen dropping resistor is *too little*. Check the value of the resistor and the value of the source voltage. If both are correct, then I_S must be low. Check the tube. If it is good, proceed with checks at other tube terminals. If screen voltage is *lower* than normal, it may be the result of low B+, a leaky screen by-pass capacitor, a defective tube, wrong bias or plate voltage, or a changed value of screen dropping resistor.

Point 4. The grid or base voltage with respect to the cathode or the emitter is very important because it is the bias that determines I_P or I_C. If the bias is *greater than normal in a vacuum tube*, I_P will be *too low*. If bias is *greater than normal (less than one volt) in a bipolar transistor*, I_C will be too high, and V_{C-E} will be too low in circuits using a load resistor. A wrong bias voltage on a transistor is such a small amount different from normal that it is best detected by checking V_C or V_E.

If bias is *less than normal in a vacuum tube*, I_P will be *too high*. This is most often caused by a leaky coupling capacitor, a "gassy" tube or trouble in the bias supply (where one is used). If bias is *less than normal in a transistor*, I_C will be *too low*. This may be due to changed value components, and sometimes bias adjustments may be set improperly. Also, I_C is a function of beta, and low I_C can be merely a result of lower beta in a transistor that has been used to replace a defective one.

Vacuum tubes are "normally on" devices; the purpose of bias is to bring I_P down to the level at which it is desired to operate. Bipolar transistors are "normally off" devices; the purpose of bias is to bring I_C up to the level at which it is desired to operate.

When bias on a power amplifier tube is very low, zero, or positive, I_P will become dangerously high. The tube itself may become red hot, the cathode resistor may burn up, the output transformer may overheat and be damaged (if something else does not fail first), or the power supply may show low B+ or fail due to the excessive load.

Point 5. When the circuit includes an emitter or cathode resistor, this is a good point to check for a clue to whether I_C or I_P is too high or too low. If V_E is *greater* than normal, it means that I_C is too high unless R_E has increased in value. If V_E is *less* than normal, it means that I_C is too low, unless R_E has decreased in value. Check bias and collector voltage for additional information and for clues as to what the circuit conditions actually are.

Point 6. The voltage difference between Points 6a and 6b depends on the value of I_C or I_P. If I_P or I_C is *higher* than normal, V_P or V_C will be *lower* than normal. If I_P or I_C is *lower* than normal, V_P or V_C will be *higher* than normal. Check for a defective tube, improper values of bias, or other static condition.

Points 7 and 8. These are grid voltages. Check for values called for and look for cause of any variation from normal.

Point 9. This point is similar to Point 5 of the power amplifier stage. Like Point 5, it gives a clue to whether I_P or I_C is too high, normal, or too low. Refer to comments given for Point 5.

Point 10. This point is similar to Point 3 of the power amplifier stage; refer to Point 3 for additional comments. Screen grid voltages are positive and should be reasonably close to the design value. The screen grid also must be held close to AC ground potential by the screen by-pass capacitor. An open capacitor will usually result in oscillation and distortion of the signal. Check by oscilloscope or by paralleling the suspected capacitor with a good one and note if trouble clears up.

2-4 Push-Pull Amplifiers

Single-ended AF power amplifiers such as those shown in Figures 2-5, 2-6, and 2-7 seem destined to become increasingly rare, giving way to push-pull circuits. The reasons are very persuasive:

1. Single-ended amplifiers must operate Class A for audio. *Class A is the least efficient class of operation.*
2. The more efficient Class B cannot be used single-ended because distortion is too high. *But a push-pull Class B stage*

2-4 Push-Pull Amplifiers 71

(which requires two transistors) can be both more efficient and have less distortion than a Class A single-ended stage.

3. AF output transformers are usually required for both tubes and transistors operating as single-ended Class A power amplifiers in the common emitter or common cathode configuration. With tubes, either single-ended or push-pull, output transformers are the standard way of matching dissimilar impedances between the final amplifier and the load for maximum transfer of power. *When transistors are connected in push-pull in the common collector configuration, direct coupling to the loudspeaker can be used and no transformer is required in a properly designed circuit.*

Push-pull circuits, then, tend increasingly to dominate in AF power amplifiers, even in the lowest power applications. These include small, portable, pocket-sized transistor receivers where Class B is a large factor in prolonging battery life.

The standard push-pull arrangement for vacuum tube circuits is shown in Figure 2-8a; there are a number of phase inversion methods; three of them are shown in Figure 2-8b.

Transistors permit a much wider variety of AF power amplifier circuits because the use of PNP and NPN transistors allows combinations not possible in vacuum tubes. (The NPN transistor resembles a pentode tube in some respects; there is no tube counterpart to the PNP transistor.) Two of the most popular basic push-pull AF power amplifier circuits using transistors are shown in Figure 2-9.

When considering Class B push-pull circuits it is important to keep in mind a few facts about them:

1. Two transistors or tubes are required. Each conducts on alternate halves of the signal cycle.

2. Each transistor in Class B conducts only for 180°, or one-half, of the signal cycle. When there is no signal, the current and conduction are near zero or very small.

3. Each transistor should be closely matched to the opposite one in push-pull; failure to do this produces imbalance and distortion.

Usually, neither tubes nor transistors are biased precisely at cut-off. Actually both draw a very small amount of current because it is standard practice to set the bias slightly toward Class A to avoid cross-over distortion.

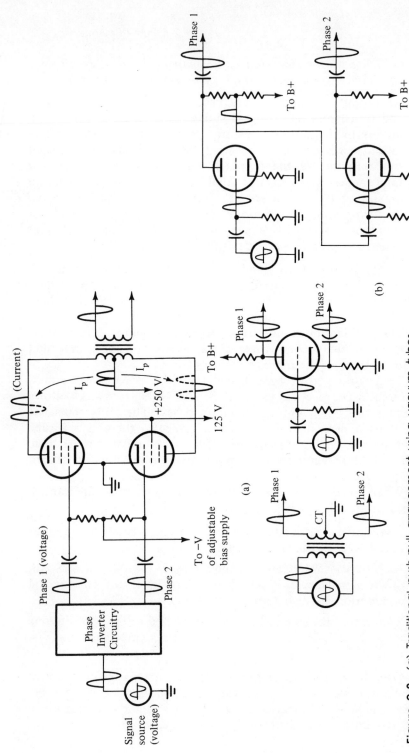

Figure 2-8 (a) Traditional push-pull arrangement using vacuum tubes. (b) Three ways of accomplishing phase inversion.

2-4 Push-Pull Amplifiers

Figure 2-9a illustrates a traditional push-pull circuit arrangement that requires a single DC power supply with R-C decoupling to provide lower voltages for other amplifiers. Note that a center-tapped AF input transformer provides phase inversion to drive the two push-pull PNP power amplifier transistors, whereas the voltage drop across forward biased diode D_1 supplies the small "turn-on" bias required by the transistors to avoid cross-over distortion. The push-pull transistors are connected in CE configuration to achieve greatest power gain; this configuration requires the use of an output transformer for impedance matching to the loudspeaker. The circuit also incorporates a negative feedback loop from the loudspeaker to the emitter of Q_1 to stabilize gain and improve fidelity.

Troubleshooting of this circuit involves sequences previously discussed to isolate faults. The troubleshooter may choose to use signal injection and/or signal tracing first to determine which stage is not operating correctly. Then, he will return to his series of static checks to determine what component or circuit has changed in value enough to affect dynamic performance.

Figure 2-9b shows a push-pull AF amplifier using a single power supply which does *not* require AF transformers or phase inversion prior to the push-pull stage. In fact, the use of a matched pair of complementary power transistors (one PNP, one NPN) accomplishes conduction on alternate half cycles when driven by a single signal. This method of achieving push-pull operation is known as complementary symmetry and is very popular. Note, however, that the push-pull transistors are connected in common collector (CC, sometimes called emitter follower) configuration; this provides no voltage gain, only current gain, hence the signal at the base must be slightly larger in voltage than the signal delivered to the loudspeaker from the emitter.

The required "turn-on" bias to avoid cross-over distortion is provided in this circuit by the voltage drop produced by I_C of Q_2 as it flows through R_1 and D_1. This voltage drop is applied across the base-emitter junction of Q_3 in series with the base-emitter junction of Q_4.

Many other variations of circuitry are found, depending on the power output required, B+ levels available, and other factors. The troubleshooter must look at each individual circuit to identify which strategies the circuit designer has employed and how he has carried them out in terms of components and hardware.

Certain elements are common to all of these circuits. These should be looked for and identified as a matter of routine: (a)

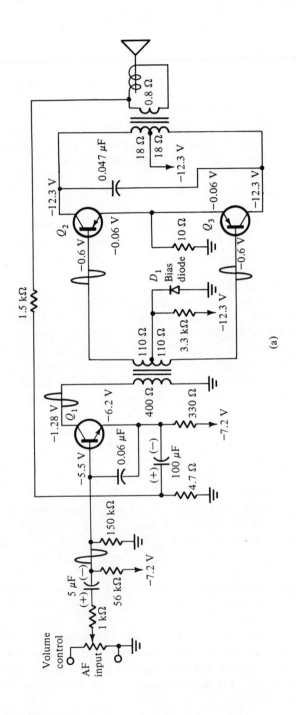

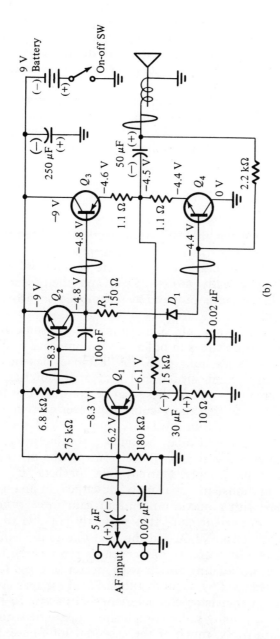

Figure 2-9 (a) Audio amplifier employing transistors and conventional transformer phase inversion and impedance matching of the push-pull amplifier to the speaker. (b) Audio amplifier using complementary symmetry in the output stage to achieve push-pull operation without requiring phase inversion preceding the output stage.

75

push-pull operation; (b) presence of "turn-on" bias to avoid crossover distortion; (c) some method of phase inversion so that the push-pull transistors conduct on alternate half cycles; (d) impedance matching to the speaker, either by the use of a transformer or by use of the CC configuration which permits coupling through a capacitor or directly to the speaker.

Effective troubleshooting of push-pull amplifiers consists of applying standard troubleshooting techniques within a background of knowing how such circuits work, and specifically "what should be there" at various points in that particular circuit.

2-5 Now, About Transistors

Since their invention in 1948, transistors have been rapidly replacing vacuum tubes in nearly all low-power applications. The attraction of transistors is that they are smaller, more efficient, longer lived, physically more rugged, and their cost has been steadily falling as mass production techniques have been refined and improved.

Practical checking of transistors differs somewhat from checking of vacuum tubes in order to accommodate to differences in the devices. A vacuum tube, for instance, depends on size, spacing, number, composition, and heat of its internal elements for its characteristics. All of these can be controlled in the manufacturing process so that the vacuum tube user can expect characteristics of most new tubes to fall within ±15% of specifications. Thus, checking vacuum tubes becomes a routine matter, using a tube checker or substitution. Transistors, on the other hand, are not always as predictable. Current gain, or beta (h_{FE}, h_{fe}) of a given type of transistor is usually described in manufacturer's specifications with a maximum current gain *three times greater than minimum current gain!* It is perfectly possible to have one transistor with a beta of 60 and the next one of the *same type* to have a beta of 180. Stability in transistors must be built into the design of the surrounding circuit with greater care and to a greater degree than required by vacuum tubes. For these reasons a definitive checker of transistor characteristics turns out to be a more expensive, and usually more complicated, instrument than a tube checker. Fortunately, much of the essential information for troubleshooting can be gained by the simpler methods which we shall now describe.

2-5 Now, About Transistors 77

The common bipolar transistor exists in two versions, NPN and PNP, which appear as diodes in an ohmmeter check, as illustrated in Figure 2-10. Also shown are a number of ohmmeter checks that are extremely useful to know. Using only an ohmmeter, we can identify the base terminal of a bipolar transistor because it is the only terminal that exhibits symmetry to the other two: We can also determine if the transistor is PNP or NPN by

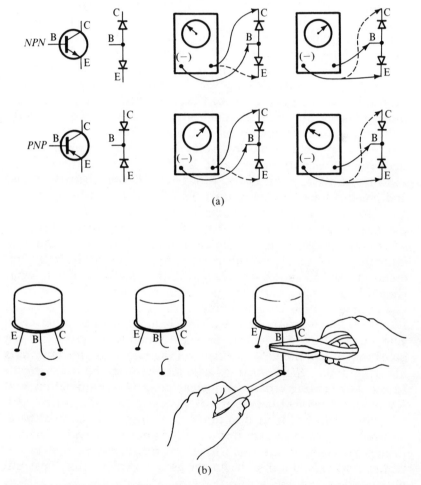

Figure 2-10 (a) Checking bipolar transistors with an ohmmeter. Note symmetry. (b) In-circuit ohmmeter checking sometimes requires disconnecting at least the base by unsoldering or cutting the lead. It is good practice to use a small soldering iron and heatsink when reconnecting.

noting when conduction is shown. It is not possible to distinguish between collector and emitter by this method. (If collector and emitter are inadvertently reversed in the circuit, gain will be very, very low; if proper, gain will be normal for the circuit.)

When checking a transistor circuit, the troubleshooter uses voltage checks in much the same fashion as for a vacuum tube circuit. Certain differences stand out, however. First, static voltages from the power supply are much higher (and more dangerous) in vacuum tube circuits. Second, signal voltages applied to the base of transistors in the CE configuration seldom approach one volt whereas signal voltages at the grid of a tube may go as high as 100 or more volts in some TV circuits (not audio). Third, transistors tend to be far less forgiving of momentary surges of excessive voltage and current than vacuum tubes; a pulse that a vacuum tube takes in its stride will destroy a transistor. Equal care must be taken not to "short out" something when working with either transistor or vacuum tube circuits, because parts will burn out just as quickly in the low-voltage transistor circuit as in the high-voltage vacuum tube circuit.

Bias in a transistor is the voltage and current betweeen base and emitter. Bias in a vacuum tube is the voltage between grid and cathode, with the grid normally negative with respect to the cathode. In an NPN transistor the base is normally 0.2 to 0.7 V *positive* with respect to the emitter; in a PNP transistor the base is normally 0.2 to 0.7 V *negative* with respect to the emitter. Reference to service notes or schematic drawings should tell what should be there in a particular piece of equipment.

Transistors fail by developing open or short circuits between terminals much more often than by changing beta or other characteristics. Usually a check of static conditions gives enough clues to point to the defect. *When voltages at transistor terminals are not normal, there must be a fault either in the transistor or in the surrounding circuit.* A logical process of elimination will reveal which one is the culprit.

One might be inclined to think that voltages as small as base-emitter voltages are insignificant. This is not true! The existence, or nonexistence, of this small base-emitter voltage is vital to circuit operation, and a troubleshooter uses it as one of his most important check points!

chapter 3

Radio frequency amplifiers: Troubleshooting and alignment of intermediate frequency amplifiers in radio and television receivers

Radio frequency amplifiers in radio and television receivers are found between the antenna and second detector in AM, between the antenna and sound demodulator in FM, and between the antenna and video detector in TV. In addition, RF amplifiers are normally used in TV between the video detector (AM) and the sound demodulator (FM) to amplify the sound carrier frequency (4.5 MHz). The amplifier has two different designations, depending on where it is placed in the signal path. If the RF amplifier precedes the mixer or converter stage, it is simply called an *RF amplifier*; if the RF amplifier follows the mixer or converter stage, it is referred to as an *intermediate frequency (IF) amplifier*. Both handle radio frequencies, and the designation RF or IF merely identifies their placement and characteristics more closely.

Typically, RF amplifiers are narrow band, frequency selective, and tunable. Nearly all radio frequency amplifiers in home electronics equipment are low level, Class A stages that usually operate in the common emitter (CE) configuration to achieve maximum power gain. The poor efficiency of Class A is not a significant deterring factor at low power levels. Besides, it is greatly outweighed by Class A's high power gain, which is of utmost importance in raising signals to usable power levels in a minimum number of amplifying stages.

Typical RF amplifiers depend on resonant circuits that are tuned to the frequency of the signal they handle to accomplish the desired selectivity and bandwidth. Resonant circuits constitute highly efficient loads for transistors and vacuum tubes when placed in their collector and plate circuits. They are almost ideally suited to transformer coupling of energy from one stage to the next. When the correct capacitor values are connected across them, they give current gain in the parallel resonant primary side and voltage gain in the series resonant secondary winding. Usually, such resonant circuits are shielded to minimize both radiation and pick-up of electric and magnetic fields that might cause oscillation, distortion, or other problems.

The objectives that RF amplifiers are expected to fulfill may be summed up as follows:

1. An RF amplifier circuit must *select the desired frequency and reject unwanted frequencies.* Resonant circuits are the usual means of accomplishing these objectives.
2. An RF amplifier must *amplify modulated signals with minimum distortion.* High gain is also desirable, so Class A is used.
3. An RF amplifier must *provide whatever bandwidth the modulated signal requires.*

Troubleshooting of RF amplifiers has one more dimension than troubleshooting of AF amplifiers: All static voltages in an RF amplifier may be perfect, yet the circuit will not perform its dynamic function satisfactorily until resonant circuits are properly tuned.

3-1 Urgent Considerations in the Use of Instruments, Service Notes, and Procedures

Probably the greatest initial handicap that faces a beginning technician is his unfamiliarity with the capabilities and limitations of the instruments he works with. Even among experienced troubleshooters not understanding instruments frequently leads to wrong conclusions, lost time, and poor work.

There are many manufacturers of test equipment intended for use with home electronic units. In addition, a particular piece of test equipment is available in several different quality and cost levels ranging from simple kits to high-grade laboratory tools.

Each instrument and quality level has its own limits and characteristics set forth in its specifications and its own set of functions, controls, layout of knobs, dials, meters, and leads. And each has subtle differences in performance that reveal themselves only through use.

The first thing anyone should do before attempting troubleshooting and alignment of RF amplifiers in radio or TV is make sure he knows how his signal generator works, how it should be connected to the unit being tested, how the function and level controls of the generator should be set, and what the output monitor of the unit under test should show when everything is working correctly. So the manufacturers' manuals should be studied and, if possible, some step-by-step practice done on a good receiver before tackling one with trouble.

What we have noted about test equipment is also true of equipment to be serviced: radios, TV receivers, and high fidelity amplifiers. With these, too, there are *many* manufacturers, quality levels, and model changes from year to year. Even the most capable person wastes time when he is confronted by a unit new to him if adequate service notes are not available. Having good service notes on the equipment to be repaired is just as important as knowing the capabilities and limitations of the test equipment to be used.

Each type and model of receiver has its own unique complement of static values of components and DC voltages, but has typical dynamic signal levels at strategic points. Each also has a recommended series of alignment adjustments, procedures and check points, plus information on what the monitor should show when checking performance by oscilloscope, voltmeter, ear, or other instrument. Service notes contain the special instructions, precautions, data, modifications and exceptions, or test procedures peculiar to a particular model.

The foregoing discussion of instruments and service notes explains why the following discussion of troubleshooting and alignment is somewhat generalized. Any discussion of alignment signal generators cannot be both specific and perfectly applicable to all available alignment equipment. Similarly, there is no way a discussion of how to align a specific receiver can apply exactly to other models. Therefore, the troubleshooter must learn the particular characteristics of his own test equipment and of the equipment he is testing.

3-2 Injecting Signals into RF Amplifiers: Practical Considerations in the Use of Signal Generators

Signal generators are indispensable in troubleshooting. They permit rapid dynamic checking of subsystems and larger units by providing the following:

1. *A choice of output signal frequency,* which are adjustable to enable the technician to match the frequency requirements of the unit or subsystem being tested.
2. *A choice of signal characteristics* — AM or FM, sine wave, or square wave — to match those of the unit being tested.
3. *A choice of output signal level.* The output level must be set to agree with the normal signal input level of the stage being tested.

Once the appropriate signal generator has been set for proper frequency and signal characteristics, the question is how to put it to use. For safety, we set up equipment as shown in Figure 3-1. (Refer to Section 1-8 and Figure 1-10 for discussion of signal generators.) The following are a few additional suggestions:

1. The ground lead of the signal generator should be connected to a convenient ground point of the equipment under test.

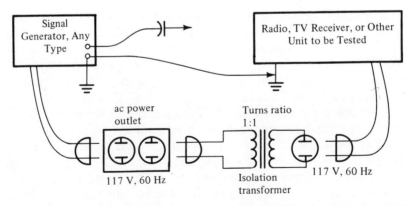

Figure 3-1 To minimize shock and damage potential, an isolation transformer is recommended for all troubleshooting. This is especially important if the equipment to be tested uses a voltage doubler or other transformerless type of DC power supply.

(Any ground point may be used, since *all* ground points in a unit are connected together, or common.)

2. The "hot" lead of the signal generator should be connected to the desired signal input point of the unit or subsystem being tested. It is recommended that a series capacitor be inserted (at about .01μF for AF and 100pF for RF) in the hot lead to remove any possibility of the signal generator upsetting the static DC conditions of the unit under test. This is a possibility because the generator represents a potential current path to ground that does not exist before the generator connection is made (see Figure 3-2).

3. Once the generator is connected and all units have power turned on, the generator should be set to the proper frequency and signal characteristics. Also set the output signal voltage of the generator just high enough to provide usable output signal from the unit being tested. This avoids distortion due to overdriving the amplifier and permits a comparison of actual gain with what should be there. More important, it permits experimental adjustment (alignment) of resonant circuits to reach optimum performance.

3-3 Intermediate Frequency Amplifiers in Radio Receivers

We follow our introduction to RF amplifiers in general by examining the characteristics of IF (intermediate frequency) amplifiers.

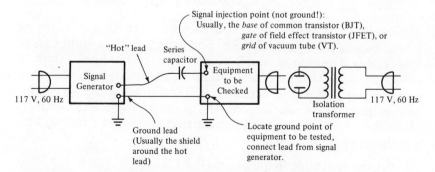

Figure 3-2 Basic equipment setup when using signal generators in dynamic operational testing, troubleshooting and alignment.

84   Radio Frequency Amplifiers

We start with IF amplifiers for a number of reasons: (1) IF amplifier signals are lower in frequency than incoming antenna signals and somewhat easier to deal with as a result. (2) IF amplifier signals are at higher power levels than antenna signals and so easier to deal with. (3) IF circuits *must* be aligned (adjusted to proper resonance) prior to dealing with circuits handling the antenna signal. (4) IF circuits, once aligned, do not have to be readjusted for long periods of time, often for the life of the equipment. Figure 3-3 shows the relationship of the IF to the rest of the receiver. Receivers that conform to this plan are called superheterodynes.

While there is nothing in theory that excludes the use of other frequencies, certain IF frequencies have become more or less standard in home equipment. Examples are:

1. For AM radio receivers operating on the standard broadcast band (from about 550 kHz to 1650 kHz), the IF is 455 or 456 kHz, or sometimes 465 kHz. Some car radios have an IF of 262 kHz.
2. FM radio receivers for the standard FM band (from about 88 MHz to 108 MHz) have an IF of 10.7 MHz.
3. For TV receivers, the picture IF operates in the 42-47 MHz region with a wide band-pass that is ideally 4 MHz. The sound IF within the TV receiver is 4.5 MHz.

To provide continuity with our previous discussion of power supplies and audio amplifiers in Chapters 1 and 2, we shall open our discussion of specific IF amplifiers with those found in the small tube-type model AM radio receiver introduced previously.

3-3a Troubleshooting IF Amplifiers in AM Receivers

Troubleshooting an IF amplifier begins *after* the AF amplifier and power supply have been checked and found to be working properly. Either dynamic or static checks may be used as a starting point, at the troubleshooter's option. In many cases it will save time to run a dynamic check immediately, using the signal injection method, before undertaking static checks. This will require the use of an amplitude modulated signal generator and an alignment tool as follows:

1. Set the generator frequency to 456 kHz (or whatever the IF of the receiver calls for).
2. Set the generator to produce a modulated signal (AM).

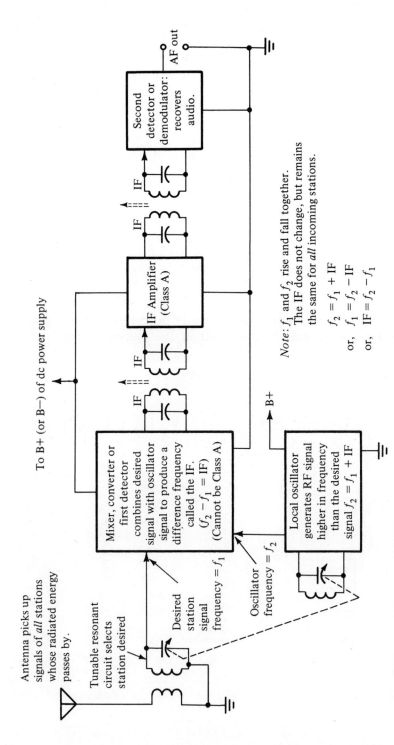

Figure 3-3 Frequency relationships between desired signal (from antenna), the local oscillator signal, and the IF amplifier signal in common superheterodyne receivers in AM, FM and TV.

86   Radio Frequency Amplifiers

3. Set the generator to lowest level of output. It can be advanced from this point if sound from the speaker is not loud enough.

4. Turn on the power to the generator and radio, turn the radio to maximum volume, and find a spot on the dial where no stations are heard. Connect the signal generator to the final IF amplifier input, usually base or grid, as suggested in Figure 3-4. If sound does not come through, adjust the *generator* output level until sound is barely audible, then rock generator frequency above and below the IF setting of its frequency dial to make sure it is centered at its loudest point. If sound comes through strongly, the IF stage is operating. If not, the circuit is either badly out of alignment or there is trouble (a static problem of some kind) in the stage. Check the alignment by using an alignment tool to turn adjustments A_1 and A_2 to reach the point of loudest sound when the generator is producing the specified IF. If alignment does not help, there is other trouble. Refer to Table 3-1 for the suggested procedure.

5. If sound came through satisfactorily in Step 4, move the signal generator lead to the input of the preceding stage (this

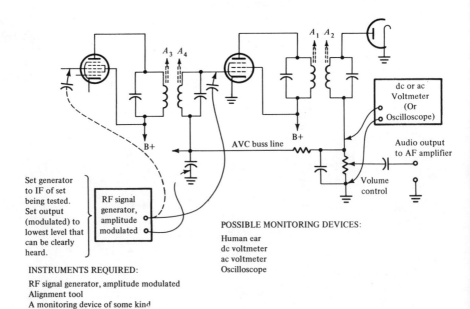

Figure 3-4 Equipment used in alignment of intermediate frequency amplifiers found in AM radio receivers.

3-3 Intermediate Frequency Amplifiers in Radio Receivers 87

may be the input base or grid of the mixer) and repeat checks of Step 4, checking adjustment of A_3 and A_4 in addition.

Before removing the signal generator lead after step 5, one should recheck all the IF tuning adjustments, aligning for maximum sound output while holding the sound at the barely audible level by *reducing the signal generator output* while the receiver's volume control is set at maximum. This insures greatest accuracy when using the human ear as the monitor. If one does not wish to rely upon the ear, a DC voltmeter across the volume control, or an AC voltmeter across the volume control or PA input or loudspeaker, or an oscilloscope across the volume control or PA input or loudspeaker will serve as more precise monitors. Take your choice.

When trouble is indicated in the preceding dynamic checks, static checks should be started to pinpoint the exact cause of poor performance. Table 3-1 is a flow chart of suggested steps. The exact sequence is unimportant, but keep in mind that checks must continue until the cause of the problem is found.

For the most part, IF amplifiers are straightforward, uncomplicated circuits that operate trouble free for long periods of time. Troubles that do occur are usually easy to identify and correct. If the steps outlined in Table 3-1 are followed, common troubles in IF amplifiers will be rapidly isolated and can be repaired with confidence.

3-3b Practical Alignment of IF Amplifiers in AM Receivers

At the present state of the art, broad bandwidth is not required in radios designed for the AM broadcast band. In fact, narrow bandwidth is welcome and sought up to the point where it begins to discriminate against sideband frequencies and AF bandwidth. For these reasons resonant circuits in AM radio receivers are generally all tuned to the *same* frequency in IF alignment.

There are different ways of accomplishing alignment, depending on how misaligned the receiver may be initially, the precision required, and the test equipment available. Although we have touched on alignment earlier in discussing how to run dynamic tests of the IF stages, we now observe that there are really three levels of alignment commonly applied to AM receivers.

First level (minimal) alignment procedure

The minimal alignment check of small AM receivers is performed when stations can be received, but only poorly. In this procedure,

TABLE 3-1

Flow Chart for Troubleshooting IF Amplifiers

Trouble: IF amplifier is weak or dead.
Possible causes:
A. A transistor or vacuum tube is defective.
B. Improper static conditions exist somewhere in the circuit.
C. Alignment is improper.

A. *Defective transistor or tube*
Checks:
1. Check tubes by testing or by substitution, then recheck operation. If normal, heat run to test performance. If trouble persists, of to B.
2. Perform static checks listed under B for a transistor first. If these suggest a bad transistor, check as described in Chapter 2, or by substitution. If transistor is good but operation is still not normal recheck B, then go to C.

B. *Improper static conditions in circuit*
Checks:
1. Measure dc voltages at device terminals and check for cause of any variations from normal which exceed 15%.
2. If dc voltages are normal there may still be a short in a coil or capacitor of a resonant circuit, or the resonant circuit may be out of alignment. If trouble persists at this point go to C and check for proper alignment.
3. If any circuit cannot be tuned to resonance, it is suspect. Check with ohmmeter and replace IF transformer if no definite cause of failure to tune properly can be found.
4. Heat run after repairs have been made to be certain that the cause of trouble has indeed been found and corrected.

C. *Improper alignment*
Checks:
1. Initiate and follow dynamic testing procedures of Section 3-2a. If this restores proper operation, heat run to test performance.
2. If dynamic checks of alignment fail to correct trouble, return to A and B for rechecks. If any resonant circuit failed to tune, go to B-3.

INSTRUMENTS:
VTVM or VOM,
AM signal generator,
Oscilloscope (optional, see text)

3-3 Intermediate Frequency Amplifiers in Radio Receivers

alignment adjustments are made for maximum sound output using the human ear as the monitor and a weak (barely audible) radio station as the signal source. This is the procedure:

1. Tune in a weak station with the radio volume advanced to maximum. Make sure the tuning is centered on the station.
2. Locate the IF transformers and determine the type of alignment tool needed to make adjustments. Best results call for the alignment tool to be nonmetallic to avoid affecting the point of resonance. Both the primary and secondary windings of many IF transformers must be adjusted. In some transformers both adjustments are accessible from the top by using a special alignment tool. In others one adjustment is accessible from the top, the other from underneath, and removal of the chassis is usually required.
3. As you listen carefully for maximum sound, rotate *each* adjustment up and down until the point of maximum volume is located. Then, move to the next adjustment and repeat the process until all adjustments have been checked.

We note again that this minimal alignment procedure requires no test instruments other than a suitable alignment tool, and that it is applied only when at least some stations can be heard.

Second level alignment procedure

If no stations are audible, or if a definite, dependable, stronger, controlled signal of known frequency is preferred, an RF signal generator may be used as the signal source instead of a radio station. If necessary, the signal may be injected into the last IF stage first, as in the troubleshooting (Figure 3-2). After the last IF stage is adjusted, the signal generator lead is moved to the input of the preceding stage for adjustment. Except for the use of the RF signal generator as a signal source, the procedure is identical with minimal first level alignment. Volume control is set at maximum, and the sound output level is kept barely audible by reducing the generator output as adjustments are made.

Third level alignment procedure

The most precise alignment requires a signal generator and an AC or DC voltmeter or oscilloscope as a monitoring device. The DC voltmeter should be connected across the volume control and set to a low voltage range. The AC voltmeter or the oscilloscope, if used instead, may be connected across the volume control, the input going to the final power amplifier or the speaker terminals. Since the final power amplifier and the speaker terminals have higher signal levels than the volume control, they may be preferred.

If the receiver still does not bring in any stations or if reception is poor after satisfactory IF alignment, and if the AF amplifier and the DC power supply have been checked and found satisfactory, the trouble will be in the oscillator, mixer or RF section preceding the IF circuits.

3-3c IF Alignment in FM Receivers

Troubleshooting and proper alignment of FM receivers follows a pattern similar to alignment in AM, but with some new considerations. First, proper alignment in FM usually requires an FM sweep generator. Second, although the ear may be used, the preferred method of monitoring the alignment process is through an oscilloscope pattern. Third, FM operates at higher RF and IF frequencies than AM and is more affected by small amounts of stray capacitance and inductance.

To insure the best reception in any type of wireless communication, the proper bandwidth must be maintained in the receiver. In FM, particularly in those receivers designed for high fidelity systems, this means that the IF response curve must be correct, including whatever type of demodulator circuit is used to recover the AF.

The response curve of any amplifier can be drawn using data obtained by a DC or AC voltmeter and a suitable signal generator, by a process of measuring the output for one frequency at a time until the whole bandpass is revealed. But a faster method is available: a sweep technique that employs an FM sweep generator and an oscilloscope. In this technique, an FM sweep generator moves the carrier frequency up and down across the desired frequency range, usually at a 60 Hz rate derived from the commercial AC power line. Thus the same information appears across the demodulator output as for the one-point-at-a-time procedure, only now it is repeated 60 times per second! More important, if an oscilloscope is used as the output monitor, the effects of any IF adjustment on the response curve can be seen immediately while it is being done.

Figure 3-5 illustrates the time relationship between the AF and the RF carrier frequency. Let us say arbitrarily that in the

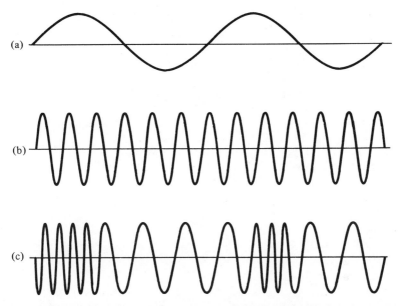

Figure 3-5 (a) The AF signal used to modulate the carrier. (b) The RF carrier signal is regular in period and frequency prior to modulation (this is the "center" or "resting" frequency in FM). (c) The FM signal after modulation deviates to higher frequencies (shorter periods) and to lower frequencies (longer periods).

transmitter a *positive-going* AF voltage (from static "no signal" conditions) causes the period of the RF carrier to be shorter, i.e., higher in frequency; and that a *negative-going* AF voltage makes the period of the RF carrier longer, i.e., lower in frequency. Thus, excursions from the center, or resting, frequency occur in step with the AF used to modulate the carrier. How far the deviation goes from the center frequency depends upon the size of the AF voltage, in other words, on the loudness.

To actually see the response curve (Figure 3-6), the output voltage of the demodulator (the AF) is connected into the vertical axis of the oscilloscope as in customary use. The spot of the oscilloscope must be made to scan horizontally across the face of the CRT (cathode ray tube) in step with the 60 Hz used to modulate the carrier. To do this, most oscilloscopes provide for *external horizontal sweep* by switch positions and terminals. Sweep generators bring the 60 Hz modulating voltage out to terminals which make it easily accessible. The operator connects the AF (60

92 Radio Frequency Amplifiers

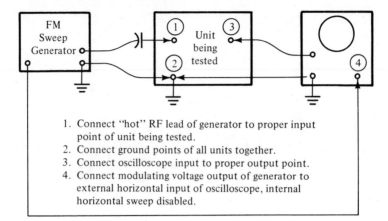

1. Connect "hot" RF lead of generator to proper input point of unit being tested.
2. Connect ground points of all units together.
3. Connect oscilloscope input to proper output point.
4. Connect modulating voltage output of generator to external horizontal input of oscilloscope, internal horizontal sweep disabled.

(a)

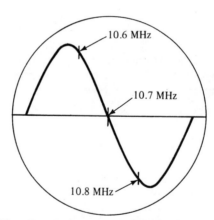

(*Note*: The 10.7 MHz marker "blip" should be in the center and the curve should be symmetrical above and below it.

(b)

Figure 3-6 (a) Equipment setup for FM alignment by sweep technique. (b) Oscilloscope display when FM receiver IF amplifiers are properly aligned. **Caution:** Refer to service notes of the receiver under test to find proper signal input and output points. Look to service notes, also, for important sequences, procedures, data, and special information about this particular receiver. **Do not guess** if you can avoid it.

3-3 Intermediate Frequency Amplifiers in Radio Receivers 93

Hz) sweep voltage output from the generator to the *horizontal input* terminals of the oscilloscope, which has been switched to external sweep position.

The pattern that is usually on the oscilloscope screen when all generator and oscilloscope controls are properly adjusted is shown in Figure 3-6b. Important points that tell frequency are the marker "blips" which serve as guidepoints in alignment. The 10.7MHz blip marks the center or resting frequency; the remainder of the display should show symmetry above and below this 10.7 MHz point. Marker blips are produced by injecting an accurate fixed-frequency, unmodulated RF carrier of the desired frequency (10.7 MHz, or whatever is needed) along with the FM signal from the generator. The FM signal sweeps up and down across the band and it produces the blip when it crosses the fixed marker signal. As long as the marker frequency is correct, the operator can tell where the center or any other frequency lies on the displayed response curve by watching the blip move as the marker frequency is changed.

If the sweep width control of the signal generator is advanced too far, the pattern will become very narrow; if it is not advanced far enough, the full bandwidth of the IF amplifiers will not be revealed. Alignment specifications of the manufacturer should be referred to for any special instructions or precautions that pertain to a particular piece of equipment.

Usually, the FM generator signal is injected into the mixer input and the IF resonant circuits are adjusted while the operator watches the effect on the curve displayed on the oscilloscope screen. In this way, he can tell immediately when the pattern is symmetrical about the resting frequency and what the bandpass characteristics are.

Typical troubles associated with IF amplifiers of FM receivers are: (1) vacuum tube or transistor failures; (2) component breakdowns causing a change in static conditions which makes proper dynamic operation unachievable; or (3) misalignment. Rarely do the IF transformers in FM or TV receivers break down. When they do, they do not tune to resonance properly when alignment is attempted. This is the best indicator of trouble in any IF transformer. *Very often, trouble in an IF transformer will not show up in static checks of resistance and DC voltages.*

FM receivers show two more differences from AM receivers, in addition to those referred to at the beginning of this section. First, the FM detector or demodulator circuit must be frequency-sensitive rather than amplitude-sensitive. Second, if the FM demodulator circuit is affected by amplitude variations in the signal presented to it, it must be preceded by a limiter stage to remove those amplitude variations. To the eye, the limiter circuit looks almost exactly like an IF amplifier stage except that voltages are different. When present, the limiter is the last IF stage immediately preceding the FM demodulator.

The FM demodulator circuit must produce the audio frequency by use of a circuit whose output changes in accordance with the signal deviation from the center or resting frequency. Several variations of circuitry are available to accomplish this. Some of the most popular are the discriminator, the ratio detector, and the gated beam detector. In practice, the limiter stage might be described as being continuously overdriven, with its input signal "slamming" it from saturation to cut-off at all times. In high-quality receivers, sometimes two limiters are used, one following the other. Many FM receivers have more IF stages than are found in AM receivers to insure adequate drive to the limiter and demodulator.

Minimal alignment procedure in FM

As with most procedures, there are shortcuts in FM receiver alignment that may be taken under certain circumstances. We have emphasized that proper and precise alignment must be done with instruments. You should be very hesitant about making any shortcut adjustments without instruments on high-quality equipment unless you are prepared to do a full alignment procedure using instruments if the short cut should be unsuccessful, or if you have produced a greater problem. However, when dealing with low-cost FM receivers, shortcut procedures taken with a little care and practice yield satisfactory results in many cases.

Procedure: Tune in an FM station and carefully center the tuning on the station, using the ear as the monitor. Next, tune each IF resonant circuit *except the secondary of the last IF output transformer to the demodulator* for maximum sound output. Do not touch the station selector dial, but leave it where you centered it at the beginning. Now, listening very carefully, adjust the secondary of the final IF transformer for the best quality of sound; that is, for minimum hum, noise, and distortion. When you have found the best point, test it by adjusting first to one side and then to the other to make sure the adjustment is centered.

3-3 Intermediate Frequency Amplifiers in Radio Receivers   95

3-3d IF Alignment in Television Receivers

We begin our discussion of TV alignment with a word of caution: *Do not attempt to use the human eye and ear as monitors in minimal alignment of the video IF amplifiers in a TV receiver.* Once these stagger-tuned bandpass amplifiers are disturbed, there is absolutely no way to regain the proper bandpass characteristics without using instruments. There are no shortcut procedures that are satisfactory.

By way of further explanation, broad bandwidth is required in the RF and IF section of TV receivers if fine picture detail is to be produced. To achieve this and at the same time yield adequate gain, the resonant circuits are stagger-tuned to different frequencies, and more IF stages are used. So if shortcuts are attempted and the resonant circuits are tuned to the same frequency, oscillation will almost inevitably result, and the specified bandwidth response cannot be regained without using instruments and the standard alignment procedure.

Figures 3-7 and 3-8 show typical IF amplifiers in tube and solid state TV receivers. There is little difference here between color and monochrome receivers, although some color sets have an additional IF amplifier stage and "trap" circuits to more closely approach an ideal bandpass.

To align the IF amplifiers, a sweep generator is connected to inject the FM signal into the mixer. If the receiver is a tube type, a good method is to connect the "hot" lead of the signal generator to an *ungrounded* tube shield over the mixer tube, as shown in Figure 3-9. If the set is a transistor type, the signal may be lightly coupled by capacity into the mixer at a test point provided, or according to manufacturer's instructions. The vertical input of the oscilloscope is connected across the load resistor of the video detector. Horizontal scan of the oscilloscope spot is accomplished as in FM alignment by setting the oscilloscope horizontal controls to external sweep position and connecting the modulating 60 Hz voltage from the generator into the horizontal input terminals of the oscilloscope.

Unlike FM, picture information in TV is amplitude modulated (AM). So a simple amplitude-sensitive diode detector is used in TV rather than the frequency-sensitive circuit used in FM. This results in a different oscilloscope display for TV, shown in Figure 3-9b. Note the appearance of the display when the generator sweep width is set too high, too low, and correctly.

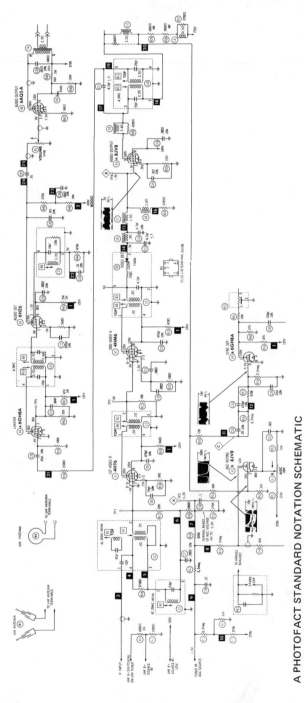

Figure 3-7 Typical video IF amplifier section in older tube type TV receiver (courtesy of Howard W. Sams & Co., Inc.).

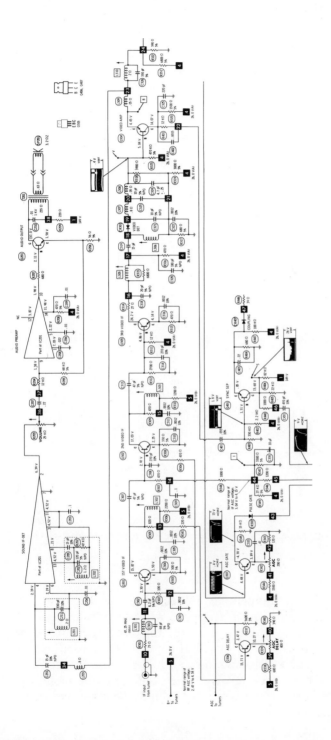

A PHOTOFACT STANDARD NOTATION SCHEMATIC
© Howard W. Sams & Co., Inc. (1974)

Figure 3-8 Video IF amplifier section in recent solid-state type TV receiver (courtesy of Howard W. Sams & Co., Inc.). Note the use of integrated circuits for the sound IF and audio pre-amplifier.

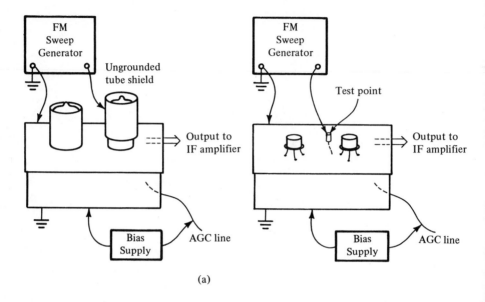

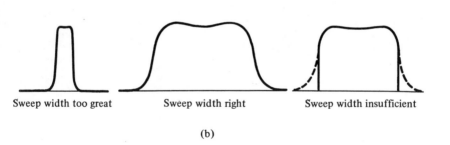

Figure 3-9 (a) Equipment setup using sweep technique for TV alignment. (b) Oscilloscope displays when connected across output of video (AM) detector.

Often the biggest problem in TV alignment is becoming familiar with what equipment is available, what settings of controls are required, and how to find or distinguish the right response curve from the spurious ones that will appear on the oscilloscope CRT screen as a result of harmonics and other interference. Once these difficulties have been resolved and some practice using the instruments has been gained, alignment should be a routine, readily accomplished procedure.

The author's experience has been that the response curve called for from a particular television receiver is often not precisely attainable in the field, even though all alignment instructions have been followed to the letter. The TV troubleshooter soon learns that when a receiver does not quite produce the exact response curve called for, it is probably due to variations in component tolerances and other factors that act cumulatively to limit performance, rather than to some undetected trouble. Much time can be wasted in attempting to attain perfection when actually it cannot be reached without redesigning the circuit.

Troubles in television IF amplifiers fall into the same three categories as troubles in IF amplifiers of AM and FM radio receivers: defective tubes or transistors, an upset in static values, and alignment problems. Actually, TV alignment problems arise rather infrequently, unless the adjustments have been tampered with or a major repair of the circuit has just been made. Resonant IF circuits in TV, like those in FM, seldom break down.

The circuitry that maintains proper voltages at the tube or transistor terminals has one particularly troublesome area. It is the AGC (automatic gain control) circuit which applies a signal-determined voltage to automatically level out the signal for strong and weak stations. Troubles here may be difficult to trace in some AGC circuits which have voltage inputs from several source points. The topic of AGC will be discussed in more detail later.

Alignment instructions of TV receivers usually call for the application of a small DC voltage to the AGC buss line. The purpose of this is to standardize the response curve as though a typical strength antenna signal were being received. To supply this small DC voltage, any well-filtered DC voltage source may be used; several manufacturers make adjustable voltage bias boxes expressly for this purpose. Service notes should be consulted to learn the magnitude and polarity of this small AGC voltage that should be applied during alignment procedures.

3-4 Summary

RF and IF amplifiers in AM, FM, and TV receivers are low-signal level, frequency-selective stages that operate Class A in the common emitter or common cathode configuration to achieve maximum power gain. These amplifiers use resonant circuit loads which must be adjusted (aligned) to the proper frequency in order for the receiver to operate correctly.

Before attempting alignment or troubleshooting RF or IF amplifiers, the troubleshooter should make sure that the DC power supply is working properly and that the AF amplifier (which follows the IF amplifiers in radio receivers) is working properly.

It is vital that the troubleshooter know how to operate his test equipment correctly, how and where to connect signal generators to the receiver being checked, how and where to connect any instrument used as a monitor to the receiver being checked, and what indication the output from the receiver will produce on the monitoring device when everything is operating right. It is equally vital that the troubleshooter know the receiver circuit, location of test points, DC voltages, typical signal levels, component values, alignment instructions, and any special information pertaining to a particular model.

Sweep alignment techniques using an FM signal generator and an oscilloscope as monitor are superior because the troubleshooter can see instantly the effect of any resonant circuit adjustment on the response curve *as it is being made.* By the use of marker frequencies he can also see where any desired frequency lies on the response curve.

Troubleshooting radio frequency amplifiers follows the same basic rules that have been discussed in the Introduction and Chapters 1 and 2. If these are followed, most common troubles will be found readily. Intermittent, rare, or obscure troubles will always be a challenge to the expert as well as the beginner, and this should be anticipated. The important point to keep in mind is, when a difficult problem is encountered, don't allow yourself to become unduly discouraged. Set it aside, do something else, and come back to it when your mind is fresh!

chapter 4

Radio frequency amplifiers: Preselectors, mixers, and oscillators in radio and television receivers

For all practical purposes, every type of radio and television receiver today conforms to the organization plan referred to as a *superheterodyne*. Key elements in a superheterodyne are:

1. A frequency selective input (usually tunable) that "chooses" the desired signal from the antenna. The signal may pass through an RF amplifier stage before going into the mixer-converter or it may go directly into the converter, omitting the RF amplifier.
2. A "local" oscillator that produces an RF signal (much stronger than the antenna signal) which also goes to the mixer-converter. The difference in frequency between the oscillator frequency and the incoming signal frequency from the antenna is the intermediate frequency, or IF.
3. A mixer-converter circuit that "mixes", "beats together", or "heterodynes" the antenna signal and the oscillator signal to produce beat frequencies, that is, the IF. To produce the beats, the mixer-converter must distort, it must be nonlinear, and it cannot be Class A.
4. With correct design and adjustment, the local oscillator will always be the same distance away in frequency from the desired antenna signal, no matter what station the unit is tuned to receive. This permits the IF stages to be "tuned and forgotten" once IF alignment is completed. Customarily, the local oscillator is *higher* in frequency than the desired antenna signal it mixes with to produce the IF.

102 Radio Frequency Amplifiers

The organization of typical AM, FM, and TV receivers is shown in Figure 4-1. In TV, the ultra high frequency (UHF) converter consists of a selective circuit tuned to the desired antenna signal, a local oscillator, and a semiconductor diode mixer. The IF that emerges from the UHF converter may then be fed directly into the IF amplifier input, or it may be fed into an unused low-frequency channel of the VHF tuner, where it is processed like any other signal on a VHF channel. This later procedure of feeding the IF into a VHF channel and mixing it a second time to produce the IF used in the receiver is sometimes referred to as *double conversion*.

4-1 Troubleshooting the Front End of a Superheterodyne Receiver

The "front end" of a superheterodyne receiver consists of the local oscillator, the mixer-converter, and the RF amplifier stage (if present) preceding the mixer. In short, the front end selects the desired antenna signal, amplifies it as necessary, mixes it with the signal from the local oscillator, and from the two signals produces the IF.

Troubleshooting the front end usually follows a sequence like this:

1. If the equipment to be checked is vacuum-tube type, check the tubes.
2. Couple loosely into the antenna an amplitude- or frequency-modulated RF signal (according to the type of receiver you are checking), and tune the receiver dial to the signal generator frequency. (Before doing this, however, test to make sure the rest of the receiver — the IF amplifiers, the AF amplifiers, and the DC power supply — are working properly.)
3. If no signal comes through, make DC voltage checks at tube and/or transistor terminals to see if proper static conditions exist.
4. If all static voltage values are normal, check resonant circuits for shorts between the plates of tuning capacitors and fixed capacitors that are a part of resonant circuits. Check alignment of resonant circuits in the RF, mixer-converter, and oscillator stages.

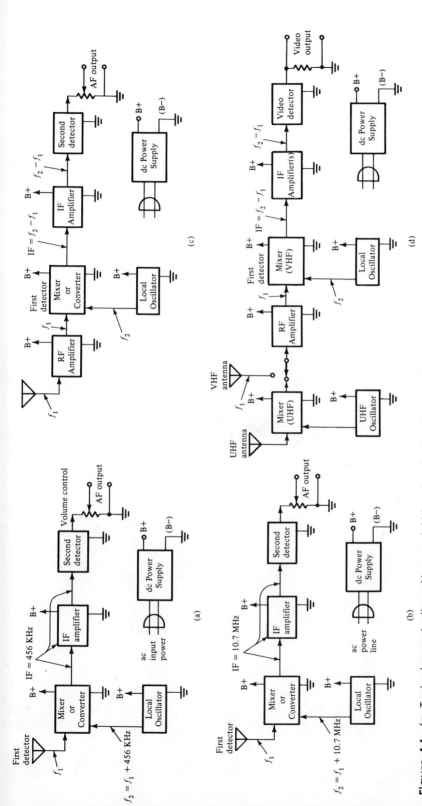

Figure 4-1 (a) Typical organization of low-cost AM radio receiver. (b) Typical organization of low-cost FM radio receiver. (c) Better quality receivers of both AM and FM usually add an RF amplifier stage. (d) Typical organization of both monochrome and color TV receivers. Note that all are superheterodynes.

When servicing AM receivers, some troubleshooters may choose an alternate approach which uses a signal tracer. The signal tracer is a high gain AF amplifier and speaker whose input comes from a detector or demodulator (diode detector) probe. Dynamic operational checking to find the defective stage using a signal tracer follows this sequence:

1. Connect the demodulator probe of the signal tracer to the signal input terminal of the first transistor or vacuum tube that the antenna signal encounters. Tune across the dial for stations (or use an AM signal generator instead of a station to furnish the signal), and listen for AF output from the signal tracer (see Figure 4-2).
2. If the signal is present at the point tested, move the demodulator probe to the input terminal of the second tube or transistor. If the signal is not there, move the probe back to the output point (the collector of a transistor or the plate of a vacuum tube) of the first stage. If there is output there, the trouble may be occurring in the coupling between stages.
3. If Steps 1 and 2 above show both the RF amplifier and mixer-converter to have input, but there is still no output from the mixer to the input of the first IF, either the mixer or the oscillator may be defective. Careful checks of DC voltages, component values, and the transistors or vacuum tubes should reveal any fault in the form of improper readings.

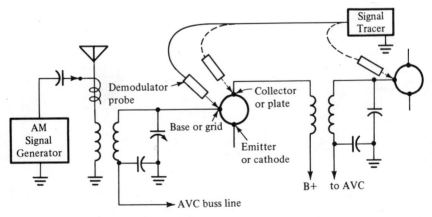

Figure 4-2 Troubleshooting AM receiver RF stages by use of a signal tracer with a demodulator probe. Note similarity to signal tracing method used in audio frequency amplifiers.

4-1 Troubleshooting the Front End of a Superheterodyne Receiver

The local oscillator is an RF amplifier with positive feedback; that is, it furnishes its own signal. If a vacuum tube is used for the oscillator, check for the existence of grid leak bias to tell if the circuit is oscillating. If the specified bias is present, the circuit is oscillating. Checking for oscillation in a transistor circuit is less certain. One way that can be used for any oscillator is shown in Figure 4-3. An unmodulated signal from a generator set to operate near the oscillator frequency is connected to the oscillator at its output point; the demodulator probe of the signal tracer is connected to the same point. Then the frequency of either the generator or the oscillator is varied through the frequency of the other. If the oscillator is working, the signal tracer will produce an audible beat note as the two frequencies are brought close to each other. If the oscillator is not working, no audible beat will be heard.

4-1a Aligning the Front End of AM Radio Receivers

To align the front end of an AM radio receiver, the RF output lead of an amplitude-modulated signal generator is connected loosely to the antenna input as shown in Figure 4-4. The procedure follows:

1. Set the modulated RF generator to the low end of the frequency band the receiver covers; set receiver dial to the same frequency. Now, if the oscillator resonant circuit has a vari-

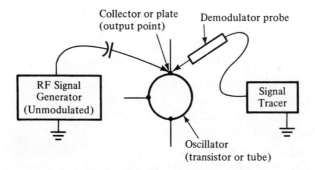

Figure 4-3 How to check to see if the oscillator is working by the beat frequency method: Inject unmodulated signal from generator at oscillator output point as shown. Vary frequency of generator through oscillator frequency, listening for a beat note as the oscillator frequency and generator frequency approach each other. If only one frequency is present no beat note will be heard.

106   Radio Frequency Amplifiers

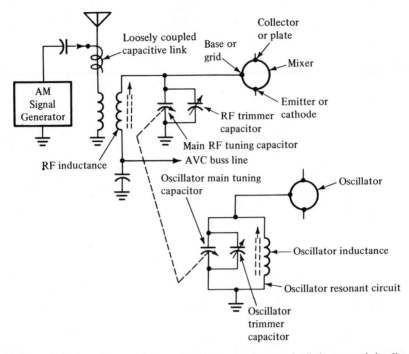

Figure 4-4 In alignment the signal generator and dial are set to the same frequency at the low end of the band and inductances of oscillator, RF amplifier, and mixer are adjusted first for maximum output. Next, both generator and dial are set to the same point at the high end of the band and trimmer capacitors are adjusted for maximum output.

able inductance, adjust the coil until the loudest sound comes from the receiver. Reduce generator output to keep the sound output at a low level as adjustments are made.

2. Without changing the generator or receiver dials, adjust the inductance (if variable) of the antenna, and any resonant circuit up to and including the mixer input, for maximum sound output from the receiver.

3. Set the dials of the RF generator and the receiver to the same frequency at the high end of the band the receiver covers. Now, adjust the oscillator trimmer capacitor to produce the loudest sound output. Do not touch inductances previously adjusted and set.

4. Without changing the generator or receiver dials, adjust the trimmers that are in parallel with the main tuning capacitors

4-1 Troubleshooting the Front End of a Superheterodyne Receiver

of resonant circuits (which handle the incoming RF signal from the antenna) for maximum sound output from the receiver.

Adjustments may be checked by repeating the foregoing sequence. At this point the alignment job is complete, and the receiver is now ready for operational testing.

4-1b Aligning the Front End of FM Radio Receivers

FM receiver alignment follows the same sequence as AM receiver alignment. The major differences are that the FM band is much higher in frequency, requiring smaller coils and smaller capacitors, and an FM signal generator is required instead of an AM. The sequence follows:

1. Couple the FM sweep generator loosely into the antenna input of the receiver.

2. Set the frequency dials of the generator and of the receiver to the same frequency at the low end of the FM band. Adjust oscillator inductance (if variable) for maximum sound output from the receiver (see Figure 4-5).

3. Without changing the frequency (that is, the dial) of either the generator or the receiver, adjust inductance (if variable) of any resonant circuit handling the antenna signal between the antenna, including the mixer input, for maximum sound output from the receiver.

4. Set the frequency dial of the generator and of the receiver to the same frequency at the high-frequency end of the FM band.

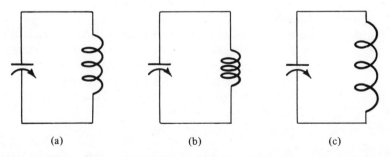

Figure 4-5 (a) Basic inductance in resonant circuit. (b) Compress coil to increase inductance and lower resonant frequency. (c) Stretch coil to decrease inductance and raise resonant frequency.

108   Radio Frequency Amplifiers

Adjust the oscillator trimmer for maximum sound output from the receiver.

5. Without changing the dial setting of either the FM generator or the receiver, adjust all trimmer capacitors of resonant circuits handling the antenna signal for maximum sound output from the receiver. This completes the front end alignment. If desired, the sequence may be repeated to make certain that all adjustments are at their optimum settings.

Some FM receivers have been manufactured with the oscillator and RF inductances consisting of a few turns of self-supporting open wire. These are intended to be adjusted by stretching out or compressing the turns, as shown in Figure 4-5.

4-2 Television Tuners

The tuner constitutes the front end of the standard TV receiver. For both monochrome and color, the most popular tuners consist of an RF amplifier stage and an oscillator-mixer. The tuner is manufactured as a subassembly designed to select the station signal desired, amplify the signal selected, mix this signal with the signal from the oscillator, and from them produce an IF in the 41-47 MHz region.

There are some general considerations to keep in mind when dealing with TV tuners: (1) Troubles stemming from dirty switch contacts and from faulty local oscillator frequency adjustment can be corrected relatively easily. (2) Vacuum tubes can be replaced easily, transistors not so easily. (3) Repair of more serious circuit troubles and mechanical troubles should be undertaken only after very careful consideration of available alternatives.

Dirty switch contacts show up by causing a snowy picture. This can be corrected in early stages of the problem by switching back and forth through the channel. (Similar symptoms may be caused by a broken antenna lead-in, but this will be evident on all channels and is not correctable by switching channels of the tuner.) Chemical contact cleaners are available in spray cans from several manufacturers which are quite effective. The cleaner is sprayed on the switch contacts and the tuner is rotated 15 to 20 times or more so that the chemical and the brushing action can combine to do their work. Unless there is a mechanical problem such as lack of spring tension or excessive wear, this procedure will restore proper operation in most cases.

Most TV tuners provide a fine tuning control for the owner's use. This varies the frequency of the local oscillator to produce the proper IF and bring in the best picture. Often, as the receiver ages the oscillator drifts until the fine tuning control lacks enough range to compensate. A wider range technician's adjustment, of which the fine tuning control is a sort of vernier, is provided to make large changes in local oscillator frequency. Access to this adjustment is usually gained by pulling off the channel selector and fine tuning knobs. This permits a working view of the oscillator end of the tuner and of the oscillator adjustment screws. Adjustment should be made with a nonmetallic screwdriver or suitable alignment tool made for the purpose.

In some tuners each channel is separate from the others. In these, each channel has its own oscillator adjustment which is independent of other channels. Other tuners utilize a series inductance arrangement for their resonant circuits, and in these *the sequence of adjustment is important.* Only by reference to service notes or through experience can one be certain which type of tuner he is dealing with and respond appropriately.

TV tuners are manufactured in various physical sizes, with varying degrees of accessibility. Some are extremely difficult to repair in the field because of their small size, hard-to-reach test points, and buried components. Then, too, any circuit repair involving the replacement of components almost always results in some change, however slight, in the placement of components, wires, and coils. This changes the distributed capacitance and tends to upset resonant circuits and amplifiers. Since small changes often have big results at these high frequencies, a tuner alignment is often desirable after component replacement.

Most large metropolitan areas have repair shops that specialize in TV tuner repair. They charge a modest flat rate, usually plus parts, for each tuner they repair. They also guarantee their work. Television dealers and repair shops, large and small, usually send tuners to one of these specialty shops for any but routine maintenance. If one is not prepared with the equipment and know-how to deal with low-level signals in the VHF range, TV tuner repair can be very frustrating and success discouragingly elusive.

The majority of today's tuner troubles, if dirty contacts are included, are mechanical in nature. Transistors have much longer potential service lives than vacuum tubes, provided their specifications are not exceeded. Transistors also operate at lower DC voltage levels, which places less stress on all components, including switches. This, too, results in longer trouble free service than vacuum tube circuits provide.

The following sequence for dealing with TV tuner problems is derived through experience over many years. It is suggested here mainly as a basis for decision making after the trouble has been localized to the TV tuner:

1. Dirty contacts and improper local oscillator adjustments are generally easy to correct. Adjusting the oscillator and cleaning switch contacts by spray chemicals are normally done on service calls.
2. Vacuum tubes are replaced on service calls. Transistor replacement is usually done in the shop rather than on a service call because they are usually soldered in and often call for a special type of transistor which has to be ordered.
3. Circuit problems involving components and mechanical problems that result from excessive wear (contacts, bearings, couplings, etc.), warped or cracked insulation, defective switch wafers or other parts are shop repair jobs. It is at this point that an assessment should be made of how to proceed before spending a lot of time: Should you attempt to make the repair yourself or should the tuner be sent out to a specialist for repair?
    a. Look carefully at size and accessibility, how many connections will have to be broken and reconnected, how many wires and components will have to be moved a bit physically, and how much mechanical disassembly and reassembly will have to be performed. If the problem is mechanical, will it require replacement of a part that will have to be ordered? Do you have the equipment to handle alignment? Do you have the time to spend if the repair should require several hours?
    b. Time is probably the overriding consideration in deciding which course to choose. If the tuner is very compact and accessibility is poor, it will take more care and time. If many connections will have to be broken and reconnected, it will take more care and time. There is also the ever present possibility of error in reconnecting a circuit or damaging some part or component in the process. If a special part has to be ordered, it may take weeks to get it—if available.

4-2 Television Tuners 111

The point of all this "negative" discussion is to caution you strongly about undertaking certain categories of TV tuner repair without adequate preparation. Unlike AM and FM receiver front ends which cover a *limited* frequency band, the TV tuner covers two *wide* frequency bands, channels 2 through 6 and channels 7 through 13. Whereas no switch contacts are required for either AM or FM bands, several switches are normal for *each channel* in TV.

Summing up, the standard servicing techniques of checking static and dynamic conditions apply equally to TV tuners as to all other amplifiers. But the special demands of TV tuner design place it in a special category of difficulty when it comes to applying these standard servicing techniques successfully.

chapter 5

Television Receivers

More than in any other type of home electronic equipment, TV troubleshooting draws upon observed symptoms that show a clear cause and effect relationship. That is, most TV problems can be localized to the defective area simply by thoughtful analysis of what is seen and heard. Insufficient picture height is clearly a vertical sweep circuit problem; not enough picture width is clearly a horizontal sweep circuit problem; poor sound and good picture is clearly a sound circuit problem; a "rolling" picture that will not lock in when both sound and picture are good is clearly a sync circuit problem; and so on. Skilled troubleshooters carefully analyze observed symptoms for a cause and effect relationship, because they know what an enormous amount of troubleshooting time this can save.

Troubleshooting TV receivers has much in common with troubleshooting other types of home electronic equipment. Like AM and FM receivers the television receiver is a superheterodyne in organization. It has an RF section, the tuner, an IF section immediately following the tuner, then a detector at the end of the IF stages. Its AF section is very much like those found in small radio receivers. Other similarities are a low voltage DC power supply to operate all of the amplifying stages and part of the picture tube, and a high voltage DC power supply to operate the remainder of the picture tube. Thus, techniques and skills used in troubleshooting audio amplifiers and AM and FM radio receivers also apply to TV.

5-1 Similarities and Differences between Television and Radio Receivers

The first major difference in TV and radio receivers is found in the different kinds of information that enter the TV receiver from the antenna. There are two RF carrier frequencies involved which carry information as described below:

1. Sound (AF) information is *frequency modulated (FM)* on one carrier which is 4.5 MHz higher than the second (picture) carrier.
2. Picture information is *amplitude modulated (AM)* on the second carrier. Color information is carried on a 3.58 MHz subcarrier and is included as a part of the picture information.
3. Vertical and horizontal sync information is *amplitude modulated* on the same carrier that has the picture information. The color sync signal or "color burst" is also amplitude modulated on the picture carrier.

All of these different kinds of information pass together through the tuner, the IF amplifier immediately following the tuner, and the video detector on their way to their respective destinations where they fulfill their assigned functions. Signal paths and termination points are shown in the functional block diagram of Figure 5-1a and b.

Unlike radio, in television repair there are video frequencies (up to 4 MHz) to deal with. In addition there is a CRT (cathode ray tube) and associated sweep and high voltage circuits which are required to make the CRT work. Finally, there is the question of color: The troubleshooter must know how black and white (monochrome) receivers differ from color receivers, and how a color signal can be received satisfactorily on a black and white set, and vice versa.

5-2 Functional Block Diagram of the Basic Television Receiver: Signal Paths and Trouble Symptoms

Anyone seriously interested in television receiver troubleshooting should begin by memorizing the functional block diagram of Figure 5-1 and the various signal paths through it. If we consider the block diagram for a moment, it quickly becomes clear how familiarity with it and with signal paths through it can speed up troubleshooting. Let us consider an example: the picture is good but there is no sound. Where do we begin to look for the cause of trouble? First, we recognize that if the trouble were in the DC power supply, the picture and the raster would also be affected. So

5-2 Functional Block Diagram of the Basic Television Receiver 115

we conclude that the DC power supply is working properly. If the trouble were in the tuner, the video IF, video detector, or video amplifier, then the picture would also be affected; we conclude that all of these circuits are functioning properly. This leaves the AF amplifier section, the sound IF amplifier, and the sound demodulator as probable sites of the trouble. If we introduce an AF signal at the volume control (refer to Chapter 2 and Figures 5-1 and 5-5), we can find out if the AF amplifier section is dead, or if the trouble lies in the sound IF or demodulator.

A second example: The sound is good and the picture tube has raster (the face of the picture tube lights up), but there is no picture. What does the block diagram tell us? First, that the DC power supply is good or there would be trouble with the sound as well as the raster. Second, that the tuner, video IF, video detector, and complete sound section are good; since the sound is taken off after the video detector, the problem must be beyond this point, in the video amplifier or the picture tube itself. Thus, we have narrowed the problem down to two areas to check for the defect just by knowing the block diagram and making thoughtful, careful observations.

If we combine the information on signal paths presented in Figure 5-1 with the list of common troubles, their symptoms, and their typical locations given in Table 5-1 we have an extremely potent weapon with which to begin troubleshooting. Table 5-1 lists troubles according to visible and audible symptoms as they appear to the viewer.

Some troubles are not caused by a breakdown of components or circuits but by gradual changes in values that finally reach the point of preventing satisfactory receiver performance. Some of these malfunctions are easily repaired because the circuit is designed to be adjusted to correct such problems. Table 5-1a presents nine of the most common TV receiver problems that can frequently be corrected by adjustments alone. Of course, the same symptoms can be caused by circuit breakdowns. But for any problem which may possibly be due to maladjustment, a basic rule is to always check the adjustments first *before* going into detailed circuit checks. If adjustment fails to correct the problem, the next step is to initiate standard troubleshooting procedures to locate the defect.

Again we emphasize that by carefully evaluating the trouble symptoms that are seen and heard, usually you can very quickly localize the trouble to a small area or subsystem on which you

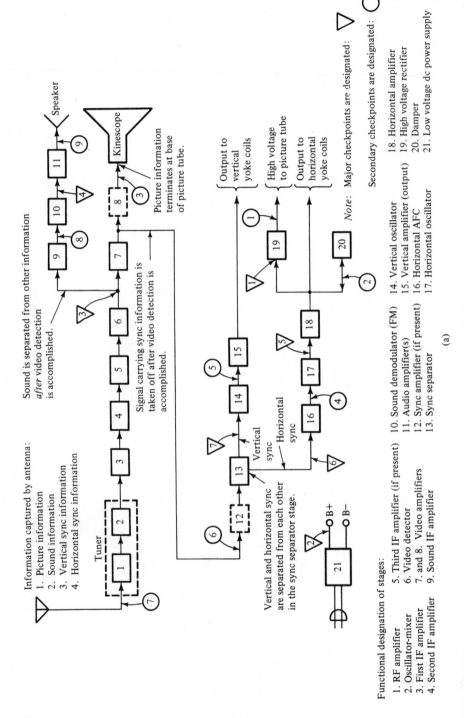

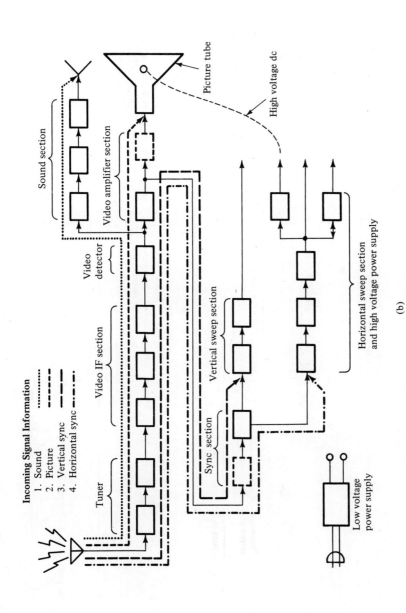

Figure 5-1 (a) Functional black diagram of typical monochrome television receiver. (b) Signal paths and termination points.

TABLE 5-1a

TV Receiver Troubles That May Be Corrected By Adjustments

Adjustment Problem	Cause and Location	Remedy
1. Fine tuning control won't adjust far enough to bring in every channel or must be readjusted for each channel.	Local oscillator adjustment wrong; usually located behind tuning knobs, or sometimes at rear of tuner.	Adjust with nonmetallic alignment tool or screwdriver in accordance with service notes.
2. Contrast too great or too little, contrast control does not correct problem.	AGC adjustment improper; may be located on rear, top or front of chassis; not always visible without removing back or covers.	Set tuner to strongest station, set contrast control to maximum, and then adjust AGC control for best picture.
3. Picture fails to lock in horizontally; horizontal hold control has insufficient range to correct the problem.	Horizontal oscillator not centered on proper frequency; adjustment may be located on top, rear, or front of chassis.	Adjust horizontal frequency adjustment to lock in picture with horizontal hold control set at midrange.
4. Picture "crowded" at top or bottom, may not fill screen at top, bottom, or both.	Vertical height and/or vertical linearity controls set improperly; controls usually (not always) located on rear of chassis.	Experimentally adjust both height and linearity controls as you watch picture; adjust to fill screen with best linearity.
5. Hum or buzz in sound; sound not clear.	a. Buzz control set improperly. b. Quadrature coil set improperly. c. Final sound IF tuned improperly. Refer to service notes for locations.	Adjust buzz control, quadrature coil, or secondary winding of last sound IF for clearest sound.
6. Picture not centered on screen.	a. Centering magnet behind deflection yoke improperly adjusted. b. Deflection yoke misaligned. Located on neck of picture tube.	a. Rotate centering magnet as you watch picture. b. Loosen yoke as needed, then align as required to center picture.
7. Picture tilted, not level.	Deflection yoke turned from proper setting.	Loosen yoke and rotate it until picture is level.
8. Picture too wide or too narrow.	Locate width control (if set has one) by reference to service notes.	Adjust as required.
9. Focus poor, picture details not clear.	Focus control set improperly; usually located at rear of chassis.	Adjust focus control for clearest, sharpest horizontal line structure and clearest picture detail.

TABLE 5-1b

Common TV Receiver Troubles, Symptoms, Locations, Checkpoints, and Procedures

Trouble	Probable Good Sections	Possible Trouble Locations	Procedures, Checkpoints, and References
1. Set dead; no sound, picture or raster.	Trouble may be anywhere; no section can be declared good on the basis of observed symptoms.	a. dc power supply b. Excessive load or short on power supply from any location in the receiver	Use VTVM initially. Check fuses und for a short across output of dc power supply. (Sound, picture, and raster seldom fail together unless power supply fails) Refer to Table 1-1 and Chapter 1.
2. No sound or picture; raster is OK.	dc power supply and both vertical and horizontal sweep circuits.	a. Tuner b. Video IF amplifiers c. Video detector d. Video amplifier(s)	Use VTVM, tube or transistor checker, or oscilloscope. If raster has snow, check tuner and antenna. If no snow, check video IF, detector and amplifier. See Chapter 3.
3. No sound; picture and raster OK.	dc power supply, tuner, video IF, detector and amplifier, sync and sweep sections.	a. AF amplifier section b. Sound IF amplifier c. Sound demodulator	Use VTVM, AF signal generator. Check AF amplifier by signal injection to see if trouble is there or preceding it. Refer to Table 2-1 and Chapters 2 and 3.
4. No picture; sound and raster OK.	dc power supply, tuner, video IF and detector, sound, sync and sweep sections.	a. Video amplifier b. Picture tube	Use VTVM, tube or transistor checker, picture tube checker, oscilloscope. Use oscilloscope to trace signal from video detector to the picture tube, make other checks as needed. Refer to Chapter 2 for techniques.
5. No picture, no raster; sound OK.	dc power supply, tuner, video IF and detector; probably video amplifier, sync and vertical sweep.	a. Trouble in horizontal sweep b. Picture tube defective c. Bias or other static voltages to picture tube are improper	Use VTVM, picture tube checker, tube or transistor checker, oscilloscope. Check for "fire" at anode of high voltage rectifier to isolate trouble. Refer to Fig. 5-2, Section 5-3, and Chapter 8.
6. Picture fills screen horizontally but not vertically.	All sections except vertical sweep section.	Vertical height and linearity adjustments, vertical oscillator, vertical yoke, output transformer or amplifier	Use VTVM, tube or transistor checker, and oscilloscope. Check for proper waveforms and proper dc voltages. Check especially for leaky capacitors. See Fig. 7-1, Chapter 7.

(continued on pages 120 and 121)

Trouble	Probable Good Sections	Possible Trouble Locations	Procedures, Checkpoints, and References
7. Picture fills screen vertically but not horizontally.	All sections except the horizontal sweep and sometimes the dc power supply	a. Horizontal amplifier b. Horizontal oscillator c. Damper, deflection yoke or flyback transformer	Use VTVM, tube or transistor checker, oscilloscope, flyback and yoke tester. Check tubes first (damper by substitution). Check horizontal amplifier drive with oscilloscope to see if oscillator is OK.
8. Picture rolls vertically and/or horizontally.	All sections except the sync; occasionally the video amplifier may compress the sync.	a. Sync separator b. Integrator network between sync separator and vertical oscillator	Use VTVM, oscilloscope, tube or transistor checker. Check tubes first, then use oscilloscope to trace sync pulse from sync separator into the oscillators. See Chapter 6.
9. Picture "tears" to the right as picture content changes.	dc power supply, sound section, vertical sweep section.	a. Improper bias on sync separator, IF amplifier or video amplifier b. Horizontal AFC or oscillator c. Improper AGC adjustment	Use oscilloscope to check for proper signal from video detector into sync separator. Use VTVM to check for proper bias and other static values.
10. Left margin of picture bends independently of picture content.	All sections except dc power supply, unless a heater-cathode short in a tube exists somewhere.	ac leakage from power line or ripple in dc power supply output is affecting timing of the horizontal sweep	Use VTVM, tube checker, oscilloscope. Check tubes for shorts and leakage. Use oscilloscope to check for ripple in output of video detector, input of sync separator and AFC circuit. Check static values with VTVM.
11. Picture is wider at top than bottom, or vice versa.	All sections except deflection yoke.	Short in deflection yoke, in horizontal portion	This condition is called "keystoning". Replace deflection yoke and check receiver for proper operation.
12. Picture does not fill screen vertically or horizontally.	All sections except dc power supply and vertical and horizontal sweep sections.	a. dc power supply b. Both vertical and horizontal sweep circuits (a coincidence)	VTVM is used to check dc power supply for low output voltage. Check static voltages in sweep circuits. Check components for changed values if power supply is OK.
13. Raster is compressed at top or bottom; sound is OK.	All sections except vertical sweep circuit.	a. Vertical height and linearity controls are out of adjustment b. Defective tube or part	Use VTVM, oscilloscope, and tube checker. Check vertical height and linearity controls first, tubes second. Check for changed value components, especially capacitors.

Trouble	Probable Good Sections	Possible Trouble Locations	Procedures, Checkpoints, and References
14. Snow in picture; sound OK or weak.	All sections except tuner, antenna, and possible AGC.	a. RF amplifier in tuner bad b. Tuner contacts dirty c. Antenna defective d. Trouble in AGC	Use tube checker, VTVM. Check tubes, spray contacts with chemical cleaner, check AGC adjustment, check dc voltages on RF stage and AGC, try antenna with another set.
15. Picture varies with sound when volume is high, is OK when low.	All except dc power supply filtering.	The dc power supply and B+ line inadequately filtered	Use VTVM, oscilloscope. Use oscilloscope to check for audio variation on the B+ line when the volume is high. Check for changed value somewhere in filter capacitor.
16. Picture enlarges and loses focus when brightness is turned up.	All except horizontal sweep and high voltage system.	Poor regulation of the high voltage power supply	Use tube checker, VTVM with high voltage probe. Check HV rectifier, damper, and horizontal amplifier. Measure high voltage with VTVM and probe for instability.
17. Picture takes a long time to come on, contrast is poor, brightness is low.	All except picture tube, low voltage dc power supply, and high voltage power supply.	a. Check picture tube b. Check high voltage supply c. Check low voltage dc power supply	Use VTVM, picture tube checker. Check CRT for low emission and shorts. Check both power supplies for proper output voltages.
18. Dark band moves slowly upward through picture.	All except dc power supply or a heater-cathode short in a tube type set.	60 or 120 Hz variation in voltage inside receiver, usually from poor filtering or from leakage	Use VTVM, oscilloscope and tube checker. Use oscilloscope to check for entry point of ripple (B+ line, output of video detector, sync separator, horizontal AFC, or oscillator).
19. Multiple images, or ghosts; picture may tear or roll.	Usually the television receiver.	Ghosts usually result from signal reflections from hills, buildings, or other solid objects; may be antenna	Try another set to be sure that the trouble is with the signal. Some cases of ghosts can be helped by using a more directional antenna or by relocating it.
20. Two bands of noise interference move slowly upward through the picture.	Almost always the TV receiver is not at fault.	This is power line (60 Hz) interference on its positive and negative peaks; it is worst in damp air and when the signal is weak	Check to see if the noise begins and stops suddenly. This will probably be some piece of electrical equipment, not the power line. If the source can be found a filter may help. This is a tantalizing problem that resists easy solution.

can concentrate your checks and measurements to find out exactly where the fault lies. This saves much time because it limits the problem to perhaps a dozen or so components as the potential trouble cause rather than to the 500 or more parts that comprise the complete television receiver.

5-3 Key Checkpoints in Television Receivers: Narrowing Down the Problem

Let us begin by noting a very important difference in the troubleshooting of radio and television receivers: It is possible to do much radio work without requiring an oscilloscope, but trying to work without an oscilloscope in television is very frustrating and a great waste of time. Proper operation of many parts of a TV receiver depends upon the peak-to-peak magnitude as well as the shape of the waveform, and the oscilloscope is the only practical instrument that can clearly verify what is actually there quickly and conveniently.

The main thrust in all types of servicing is to narrow down and localize the problem as quickly as possible. In TV this calls for the following sequence:

1. Look carefully at all available symptoms: visible, audible, odor, heat, action of switches and controls, and whether or not the problem is intermittent.

2. From an analysis of the observed symptoms, make an estimate of where the trouble lies and which subsystem(s) of the receiver may be involved. To do this you must draw heavily upon the information contained in Figure 5-1 and Table 5-1.

3. Check adjustments and controls first (see Table 5-1a), and vacuum tubes second.

4. Using the oscilloscope and voltmeter, look at appropriate key check points (see Figures 5-1 to 5-7) to confirm or refute your diagnosis. Refer to service notes to find out exactly what is specified.

5. As soon as the faulty subsystem is definitely identified, proceed with detailed checks described here and in other chapters to locate the defect itself.

There are seven major checkpoints (or test points) suggested in Figure 5-1, and nine secondary ones to further narrow the prob-

5-3 Key Checkpoints in Television Receivers   123

lem. We shall now look, one at a time, at the waveforms and values typically found at each of these major checkpoints. Keep in mind that for a given make and model of receiver one must consult service notes to get accurate and precise information about that particular set.

Checkpoint 1 is used to check for high voltage input to the high voltage rectifier when the picture tube fails to light up. If there is no high voltage, there will be no raster. This is a quick check to determine if the horizontal oscillator and the horizontal amplifier are working. The troubleshooter turns to this checkpoint when the symptoms described under Items 5, 16, and 17 in Table 5-1 are encountered.

This checkpoint is located at the anode of the high voltage rectifier diode. To make the check, the technician chooses a screwdriver with a well-insulated handle and brings the tip of it near the anode connection point after the TV receiver is turned on and has had time to warm up (see Figure 5-2). If the horizontal oscillator and horizontal amplifier are working properly, an arc of approximately ¼ inch can be drawn. Exactly how long an arc should be established when this check for "fire" is made depends upon the amount of high voltage the system is designed to produce, which in turn depends on the size of the picture tube and its requirements.

If the arc that can be drawn is too short, break the connection to the rectifier and recheck, as shown in Figure 5-2b. If the arc is now normal length, look for a defective diode or a short somewhere in the high voltage DC lead that goes from the cathode of the high voltage diode to the side of the picture tube. The fact that the arc is normal length when the load is disconnected indicates that the load is drawing excessive current, or, less likely, that the source resistance of the horizontal amplifier system has increased greatly.

If the arc is too short regardless of whether the anode is connected or not, something is wrong with the horizontal oscillator, horizontal amplifier, damper, or their circuits. Now is the time to use Checkpoint 5 to determine if the horizontal oscillator is working as it should. If it is, check the horizontal amplifier. Especially with vacuum tubes, both the horizontal amplifier and damper should be checked by substitution if possible, because no tube checker duplicates the high voltage conditions under which these tubes work in the circuit. Either tube may show "good" in the tube checker but may break down and not work in the cir-

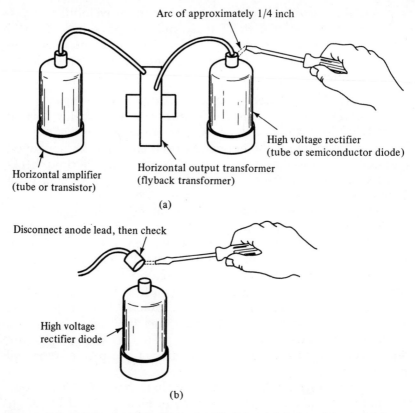

Figure 5-2 Method of checking for "fire" to determine if horizontal oscillator and horizontal amplifier are working. (a) Check first with HV rectifier connected. If arc is normal, the horizontal amplifier and oscillator are working. (b) If arc is too small in previous check, disconnect load and recheck. A normal arc now indicates possible short in rectifier diode or load.

cuit. A failure of the horizontal oscillator, the horizontal amplifier, the high voltage rectifier, or the damper will cause loss of high voltage and with it loss of raster. Other, less frequent causes of raster loss are failure of the picture tube or loss of DC voltages applied at the picture tube socket to the electron gun, an essential portion of the picture tube.

Checkpoint 2 comes in for examination when a power supply problem is suspected. We recall that the B+ output of the DC power supply must meet the following requirements if the load it serves is to work satisfactorily:

5-3 Key Checkpoints in Television Receivers

1. It must maintain the proper value of DC voltage when the load is connected and operating.
2. It must be within the specified limits of peak-to-peak ripple under normal load.
3. The B+ line must not show significant amounts of audio frequency or other variation above the normal limits of AC ripple that are called for in specifications and service notes.

Use a voltmeter and an oscilloscope to check for "what should be there" at Checkpoint 2. Refer to service notes, Figure 5-3, and Chapter 1 for guidance if trouble in the power supply is indicated.

Low DC output from the power supply in TV usually shows up first as a failure of the picture to fill the screen. It may also cause a loss of focus and occasionally may cause the local oscillator in the tuner to fail; if this happens, the symptoms are loss of both sound and picture.

Inadequate filtering in TV can result in horizontal displacement of the raster and picture elements (Items 9 and 10, Table 5-1), sound bars in the picture (Item 15, Table 5-1), vertical roll (Item 8, Table 5-1), hum in the sound, and/or a dark band in the picture (Item 18, Table 5-1).

Higher than normal ripple and loss of filtering often indicate either a drop in the electrical size of a filter capacitor or excessive load on the power supply. It may also mean that the power supply impedance as seen by the load is higher than normal. Higher than normal ripple and low B+ voltage often occur together.

We remind ourselves again, for emphasis, that the DC power supply *must* operate properly before the amplifiers to which it supplies power can operate satisfactorily.

Checkpoint 3 is used to check the output of the tuner, video IF amplifiers, and video detector when there is raster but no sound or picture. It is used when symptoms are those of Items 2, 4, 8, 9, 10, 17 and 18 in Table 5-1.

Use the oscilloscope to check at the designated output point of the video detector for proper waveform and peak-to-peak voltage for that particular make and model receiver. Refer to service notes and Figure 5-4 for guidance and information about what is called for.

If there is no output from the video detector with power on and signal input from the antenna, the fault may be in the RF amplifier, the oscillator-mixer, the IF amplifier stages, or the video

126   Television Receivers

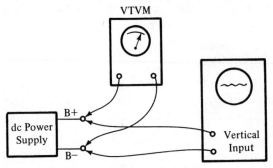

(B– is usually, but not always, grounded.
Refer to service notes if there is doubt.)

(a)

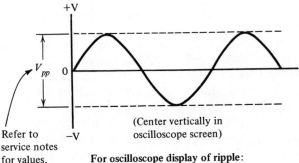

Refer to
service notes
for values.

(Center vertically in
oscilloscope screen)

For oscilloscope display of ripple:
1. Set oscilloscope for ac input to vertical.
2. Set horizontal frequency to 30 Hz. This should show 2 cycles for half-wave rectifiers and 4 cycles for full-wave rectifiers.
3. Measure V_{pp} of ripple and compare with service notes.

(b)

Figure 5-3 (a) Equipment setup for measuring DC voltage and ripple at output of power supply. (b) Oscilloscope settings and display for half-wave rectifiers.

detector itself. If the receiver is tube type, check all tubes first. If the receiver is transistorized, or if the tubes have been checked and found to be good, the next step is to check for proper static conditions on all stages that might cause the trouble. Use a voltmeter to check for correct DC voltages on collectors, emitters, and bases of transistors and on plates, grids, and cathodes of vac-

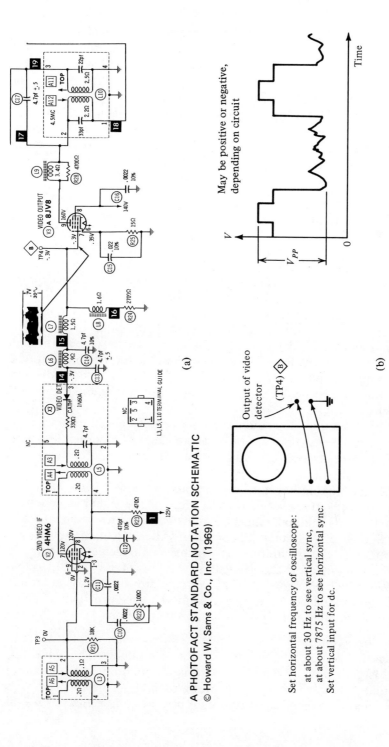

Figure 5-4 (a) Portion of TV schematic showing output of video detector and location of testpoint (courtesy of Howard W. Sams & Co., Inc.). (b) Typical equipment setup and oscilloscope display of horizontal sync. Always refer to service notes for exact information.

uum tubes. Use an ohmmeter to check for open circuits, short circuits and/or changed value components. Refer to Chapters 3 and 4 for suggestions concerning procedures and sequences. *Do not attempt to correct the problem by moving any tuning slugs or adjustments in the tuner or IF amplifier strip at this time!*

If checks so far have not localized the problem, the next step is to inject a signal (using either signal tracing or signal injection technique) and try to pinpoint the stage where the signal is being interrupted. Once the faulty stage has been located, the troubleshooter can zero in on exactly what went wrong.

Checkpoint 4 is used when there is picture but no sound. It is located at the input of the AF amplifier, usually the volume control. Refer to Chapter 2 and Figure 2-4 for suggestions about how to check an AF amplifier. Figure 5-5 illustrates the use of an oscilloscope as a signal tracer.

If sound is heard when an AF signal is injected at Checkpoint 4, it means that the AF amplifier is good and the trouble lies between the video detector and the checkpoint. The trouble may be in the sound take-off circuit, the sound IF amplifier, or the sound demodulator (FM detector) circuit. If no sound comes through when a signal is injected into Checkpoint 4, the trouble is in the AF amplifier section. Use the oscilloscope to trace the audio signal from point to point after it emerges from the sound demodulator circuit.

If the sound demodulator is suspected, check the vacuum tubes first. If the tubes are good, make checks for correct DC voltages, correct circuits (i.e., not open or shorted), and correct component values.

If the trouble has not been isolated by now, it may be necessary to check the malfunctioning area by the signal injection or signal tracing method, one stage at a time, for a few more clues.

Sync buzz or hum and poor audio will result if the tuning adjustment of the quadrature coil (gated beam detector circuit) is not just right, or if the secondary winding of the last IF transformer which goes into the FM demodulator circuit is not tuned just right. Refer to service notes on the receiver that is being checked for alignment procedures and instructions.

Checkpoint 5 is used when there is no raster or high voltage, indicating trouble in the horizontal sweep circuit. The waveform at this point reveals whether there is output from the horizontal oscillator, and whether it is working properly. This checkpoint is located at the signal input terminal, base or grid, of the horizontal

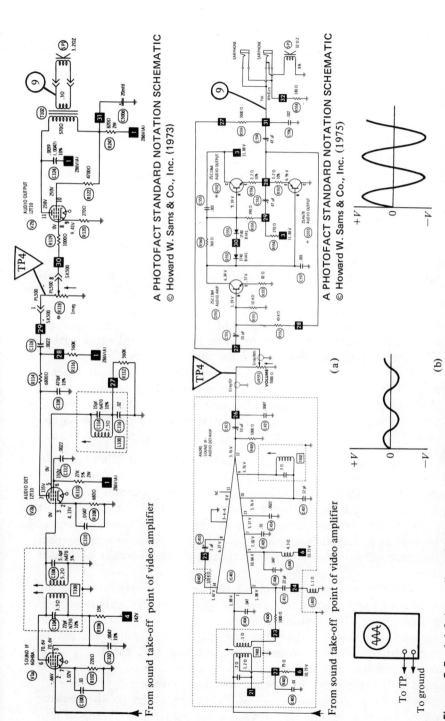

Figure 5-5 (a) Schematics of two audio sections, one vacuum tube and one solid state, with testpoints keyed to Figure 5-1 (courtesy of Howard W. Sams & Co., Inc.). (b) Signals of TP4 and testpoint 9 illustrate voltage gain.

amplifier. Whenever insufficient high voltage is found at Checkpoint 1, the next step is to use Checkpoint 5 to determine if the trouble is in the horizontal oscillator or in the horizontal amplifier circuit which includes the damper.

Use the oscilloscope to check for proper waveform and peak-to-peak voltage as indicated in service notes and Figure 5-6. If these are correct, it means that the horizontal oscillator is working properly and the trouble is in the horizontal amplifier, the damper, or their associated circuitry, the flyback transformer or deflection yoke. *Caution:* High voltages at measuring points may damage instruments. Such points are usually called out on the schematic by the notation, "DO NOT MEASURE."

Checkpoint 6 is used to check for proper horizontal sync pulse output into the horizontal AFC circuit from the sync separator. We check this point when the picture does not lock in horizontally. If the sync pulse is not proper, adjusting the horizontal hold control is ineffective. If the sync pulse is missing or smaller than service notes indicate it should be, check the sync separator for proper operation. If the sync pulse is correct and the horizontal sweep is still unstable, look for trouble in the AFC circuit.

Figure 5-7 shows the typical oscilloscope display at Checkpoint 6. By observing the pulse at this point the troubleshooter is able to determine if the fault lies somewhere in the sync separator prior to this test point or after it somewhere in the AFC circuit.

Checkpoint 7 does for the vertical sync what Checkpoint 6 does for the horizontal sync, that is, it tells the troubleshooter if the sync separator is functioning normally or whether he should look to the vertical oscillator for the fault. Note that AFC is not used in the vertical circuit as in the horizontal; instead, vertical lock-in is accomplished solely by the sync pulse emerging from the vertical integrator network to trigger the vertical oscillator directly.

Use the oscilloscope to check for the proper sync pulse at the output of the vertical integrator network. But note this most important step: *The vertical oscillator must be disabled in order to see the sync pulse.* Otherwise, the much larger pulse created by the vertical oscillator when it is operating will mask the sync pulse, and the troubleshooter will be unable to tell whether the sync pulse is correct or not. Instructions concerning the recommended way to disable the oscillator are given in the service notes. Usually, it calls for bypassing the signal at the vertical am-

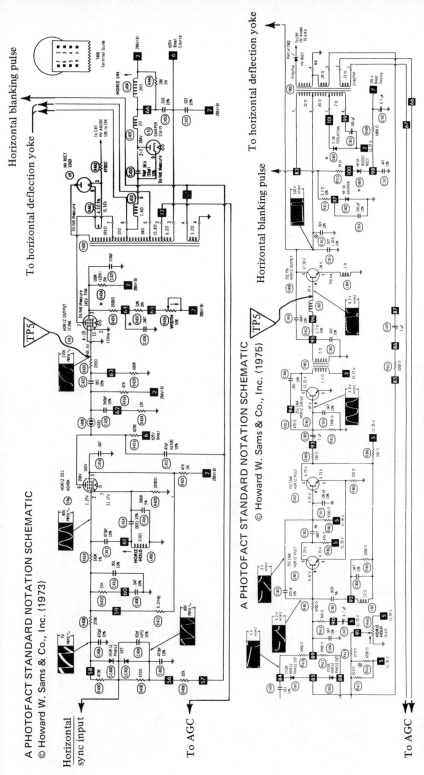

Figure 5-6 Note oscilloscope display of waveform at testpoint 5 (courtesy of Howard W. Sams & Co., Inc.) These are typical wave shapes although the peak-to-peak voltages will vary from one receiver to another. Refer to service notes to be certain.

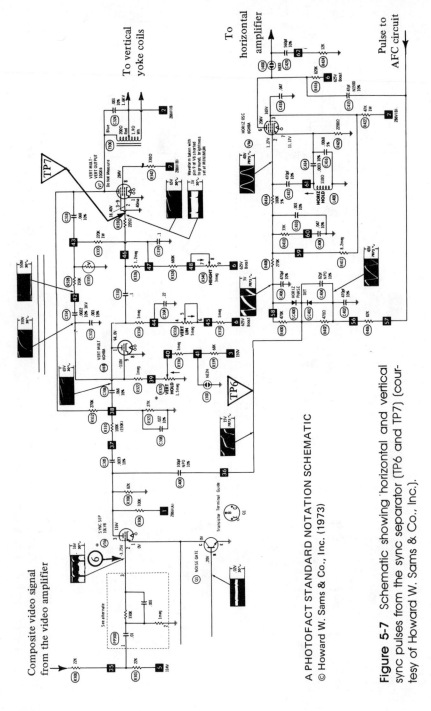

Figure 5-7 Schematic showing horizontal and vertical sync pulses from the sync separator (TP6 and TP7) (courtesy of Howard W. Sams & Co., Inc.).

plifier base or grid to ground, usually via a capacitor but sometimes directly.

Refer to Fig. 5-7 for a typical oscilloscope display of the vertical sync pulse at Checkpoint 7 when the vertical oscillator is not operating.

We have pointed out seven major checkpoints that greatly speed TV servicing including typical oscilloscope displays that are found at each one. Note that all of these are dynamic checkpoints, used to find out which subsystems are carrying out their signal handling mission successfully and which are not. As soon as the faulty subsystem is definitely identified, the secondary dynamic checkpoints can be used to further narrow down the problem. After dynamic checks have revealed the malfunctioning subsystem or stage, the troubleshooter returns to standard static checks of DC voltage and circuit values to pinpoint the defective part.

5-4 Locating Checkpoints and Other Practical Problems

A good rule to follow is always do as much troubleshooting as possible without major disassembly, such as removing the chassis. It is surprising how much can be done and how much data can be gained by ingenious use of top side (accessible) test points with only the back of the receiver removed. In tube type receivers, extension sockets can be plugged in which make the bottom terminals of sockets available for static measurments from top side.

A few general observations are appropriate at this point. First, the smaller the TV receiver is, the more difficult it is to locate and reach desired test points. In general, in older, larger, tube type receivers the circuits are more accessible and the checkpoints easier to locate and reach, although the set will require larger bench space. Second, there is almost always a problem of interconnections whenever the chassis must be removed. Speaker leads, deflection yoke leads, the high voltage lead, and the leads to the picture tube socket are usually too short to make connections when the chassis is out of the cabinet. In order to operate the receiver, extensions must be added to each of the too-short leads. Such extensions are commercially made and can be purchased from wholesalers. If preferred, they can be constructed by the troubleshooter as the need arises, although this is not an economical use of time. Extension leads are also sometimes required for subassembly work.

Some receiver subassemblies are constructed in such a way that they must be disassembled for testing, necessitating making nonstandard extension connections for B+, signal, AGC, etc., before checks can be made. Usually this problem is most severe in the smallest portables, in which all of the functional stages of large sets must be crammed into a limited amount of space.

In preparing for testing after the chassis has been removed, one must take particular care to avoid shorts between leads, with special attention given to the high voltage lead which goes to the side of the CRT. This lead carries several thousand volts and adding an extension to it poses both a potential shock and damage hazard to anyone who is careless. Unless heavy insulation surrounds the lead, trouble is almost certain.

Another problem in locating checkpoints is the possibility of physical damage—scratches on the picture tube face, plastic mask or cabinet, and breakage of plastic knobs or other fragile parts. Many troubleshooters put a tightly woven carpet on their work bench surface to lessen the possibility of damage while they work.

The job of locating components and checkpoints is really a matter of relating the actual physical placement of the component or checkpoint on the chassis to its functional electrical location on block diagrams and schematic drawings. Well-documented service notes, such as Howard W. Sams' *Photofacts* and those of some manufacturers, number each component and put a "flag" on important test points on the schematic drawing. They also give parts values and DC voltages at strategic points as well as waveforms and peak-to-peak values where they are significant. Then, on a photograph or drawing of the actual chassis the location of components and test points is called out, as indicated in Figure 5-8. By using these aids, a troubleshooter who suspects a fault in a particular subsystem from the observed symptoms is able to locate quickly both test points and components on the chassis itself.

As an example, let us imagine that a TV receiver has raster and sound but no picture. From the symptoms, we deduce that the trouble is most likely to be in the video amplifier or possibly the picture tube. This suggests that major checkpoint 3 and secondary checkpoint 3 should be checked with the oscilloscope after the tubes have been tested (if the receiver is tube type). But

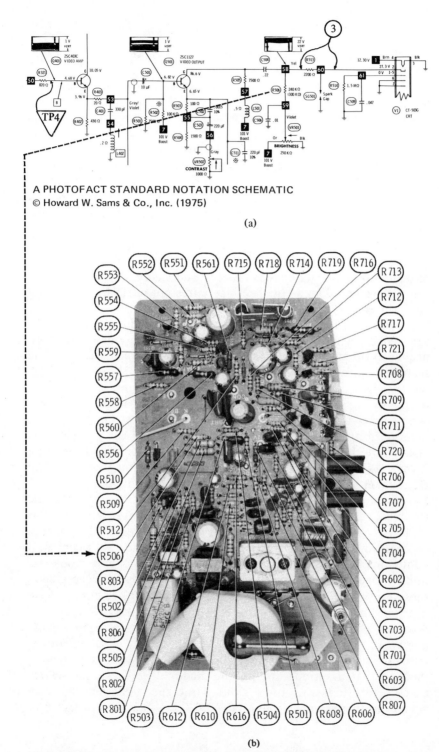

Figure 5-8 Locating parts physically on the chassis that are shown electrically on the schematic (courtesy of Howard W. Sams & Co., Inc.).

where are the video amplifier and the desired test points located on the chassis?

First, look at the schematic drawing or functional placement of subsystems given in the service notes (Figure 5-8a). Locate the video amplifier and the checkpoints on the drawings and note how they are designated. Next, look at the photographs or drawings of the actual chassis and locate physically where the desired points and components are located (Figure 5-8b). Now, look at the chassis itself and find the points sought. You are now ready to connect your voltmeter or oscilloscope, turn the receiver on, and check for "what should be there." Refer to Figure 5-8 and service notes for the essential information.

Well-documented service notes and a little practice in following the procedures just described will enable the troubleshooter to find any point on the chassis quickly and easily. But a sad fact of life is that well-documented service notes are not always at hand. Sometimes only the schematic drawing is available. In this case, the technician looks first on the schematic for the point he is interested in. When he finds it he notes what components are connected to the point of interest and what their values are. Next, he searches through the circuit *on the chassis* for the point which has the right components and the right values connected to it. Referring to the schematic, he checks points at the other ends of the components to see that they too agree with the schematic. When he is certain that the point on the schematic and the point on the chassis are the same one he is ready to make checks and measurements. As one might expect, this process of using the schematic alone takes longer because of less adequate documentation.

When neither schematics nor service notes are available, the troubleshooter has only experience to draw upon. In general, it is fairly easy to remember the typical limits between which values of peak-to-peak voltage and waveforms fall at the major test points. In the absence of a schematic one can recognize video amplifier stages by the peaking coils or ferrite beads between stages; these are found nowhere else in the receiver. Other circuits also have distinctive characteristics which make them recognizable. So although it is possible, it simply takes longer to locate checkpoints. In the event signal tracing and standard checking procedures do not succeed and the trouble still persists, it is recommended that a serious attempt be made to get a set of service notes before investing more time. Time spent guessing at what should be there is not productive, and there is always the chance that you may do something that creates a new and perhaps larger problem.

5-5 Summary

Chapter 5 seeks to do for TV what Chapters 1, 2, 3, and 4 try to do for the beginner in audio, AM radio, FM radio, and alignment of RF amplifiers. (Note that the TV tuner and the video IF strip are discussed under the topic *RF amplifiers and oscillators* covered in Chapters 3 and 4.)

Television receivers provide a unique opportunity to use trouble symptoms to localize problems to the faulty subsystem. Even without much practice, anyone who understands the functional block diagram and signal paths given in Figure 5-1 can diagnose many troubles correctly to the faulty subsystem. By also using Table 5-1, which lists trouble symptoms, probable trouble locations, probable "good" subsystems, as well as instruments, techniques, and check points, even someone new to TV can locate and repair many common TV troubles.

chapter 6

Synchronization and automatic gain control in television receivers

Because of the manner in which a television picture is transmitted and reproduced, which treats highlights and shadows as voltage versus time, synchronization of the camera and reproducer is vital in order to create a meaningful picture. Automatic gain control on the other hand is not vital, but is a convenience which allows the receiver to accommodate to a wide range of signal strengths without manual adjustment. In this chapter we examine the details of both synchronization and automatic gain control in order to facilitate the troubleshooting process.

The monochrome television picture in the receiver is made by a single spot of light, approximately the size of a pencil lead, which moves rapidly to form a lighted rectangle with a ratio of 4 (width) to 3 (height). The rectangle of light is referred to as the *raster*, the 4 to 3 ratio of width to height is called the *aspect ratio*, and the movement of the spot of light is described as *scanning*.

The spot of light is created when a beam of electrons from the cathode at the socket end of the picture tube is guided to strike the phosphor coating on the inside of the picture tube face, causing it to glow. The electron beam is deflected (moved) both horizontally and vertically by magnetic fields. These magnetic fields are produced by current through coils of the deflection yoke; one set is driven by the vertical sweep system, the other by the horizontal sweep system. Each system is separate and independent of the other, but both operate from a common DC power supply.

The critical requirement to create a satisfactory picture in the TV receiver is that the picture information carried by the moving spot be deposited in *exactly* the right place on the lighted screen at all times. This is the reason why synchronization is vital, an absolute requirement in order to "time" the spot movement properly. For example, a bright area in the upper left-hand corner of the picture seen by the camera must appear in the upper left-hand corner of the picture produced by the moving spot in the receiver. In the following pages we will examine how the picture is made, the basics of synchronization, typical troubles, symptoms of improper synchronization, and strategies for dealing with sync-related problems.

6-1 How the Television Receiver Creates a Picture

The scanning frequencies of 15,750 Hz for the horizontal and 60 Hz for the vertical arose in the early days of television when engineering standards were made. It was decided to send 30 pictures or frames per second and to scan each picture twice, making the frequency of scanning 60 Hz. Thus, 60 Hz became the vertical scanning frequency. Because this coincides with the AC power line frequency, any interference from the power line will stand still, and therefore be less noticeable than if it drifted through the picture. (This is also a useful servicing clue on color signals; 60 Hz power-line inspired interference, or poor filtering, causes the interference band to drift upward through the picture rather than stand still. See Items 5 through 9, 10, 18, and 20 in Table 5-1b.)

It was also decided to allocate 525 horizontal scanning lines to each picture in the U.S. system, as this was deemed sufficient to give adequate picture detail and quality. This point has been much debated over the years as screen sizes have grown larger but, once decided, it is very difficult and disruptive to change. There seems little likelihood of much change in existing standards in the foreseeable future.

If we multiply 30 pictures per second times 525 lines per picture, we arrive at 15,750 lines per second, the horizontal sweep frequency. This means that the time duration, or period, of one horizontal cycle is 1/15750 second, or approximately 63.5 microseconds. Figure 6-1 illustrates how the spot moves across the face of the picture tube (trace time), then rapidly moves back (retrace time) to begin the next horizontal line. The transmitter sends out a "blanking signal" to cut off the picture tube during retrace so that only the trace time is visible. Note that Figure 6-1c repre-

6-1 How the Television Receiver Creates a Picture 141

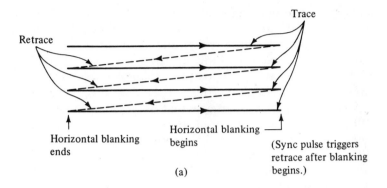
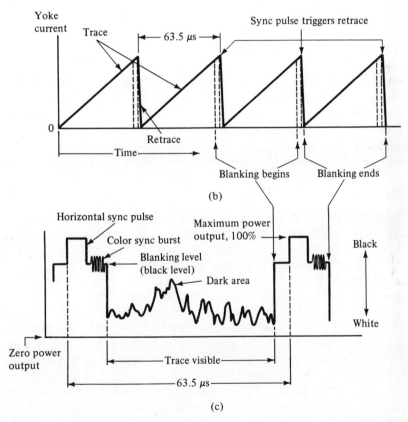

Figure 6-1 Basic details of horizontal sweep. (a) Spot moves across face of picture tube to produce raster. Note that successive traces fall below previous ones due to action of the vertical sweep system. (b) Basic timing relationships. (c) Power output of transmitter in relation to black/white level seen by viewer.

sents one cycle of horizontal sweep, such as would exist at major checkpoint 3 discussed in Section 5-3 and shown in Figure 5-1.

The truth is that the troubleshooter seldom sees a *single* horizontal sweep when looking at the video signal; ordinary oscilloscopes display a composite of *many* sweep lines and the information that each carries. This happens because of the persistence time of an image that is a characteristic of the oscilloscope CRT phosphor. (The term *persistence* refers to how long an image, once created, remains visible.) When using an oscilloscope to view the composite video signal, this simply means that the information contained in the *first* line does not have time to decay to invisibility before more lines are added on top of it. Thus, what the troubleshooter sees on the oscilloscope CRT is the sum total of many lines, not just one. However, fixed levels of voltage, such as those of blanking and sync pulses, produce steady displays because each subsequent display falls on top of the previous one. It is the video information that varies from line to line, so the trace portion of the oscilloscope display is usually indistinct and fuzzy.

6-2 Viewing the Horizontal Sync Pulse: A "Square Wave"

There is no such thing as a square wave! In electronics, the term square wave is used to describe voltage, current, or power changes that have the general appearance of a square or rectangle to the eye when the amount is plotted against time. Refer to Figure 6-2 for sketches showing how a theoretical, true square wave compares with actual square waves found in electronics.

But let us think for a moment about what the term *square wave* implies. To produce a true square wave with 90° corners, the voltage would have to change from value A to value B with no elapsed time! This cannot be done; some time, however small, is required for the voltage to change from Level A to Level B, and vice versa. The real-world waveform, then, cannot be square, but requires a certain amount of *rise time* and *fall time*, as Figure 6-2 illustrates. The horizontal sync pulse is an example of a so-called square wave. In Figure 6-1c, we note standards the horizontal sync pulse must meet.

There is another problem that the practical troubleshooter must meet: How confident can he be that the waveform he sees

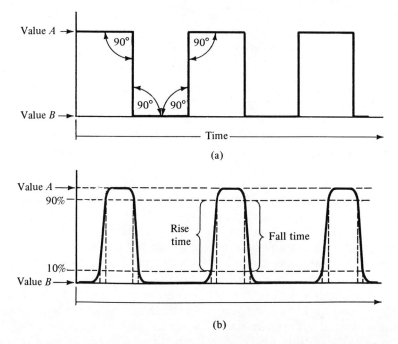

Figure 6-2 (a) Theoretical "true" square wave. Note that it has 90° angles and that it changes from Value **A** to Value **B** with no elapsed time. (b) Actual, "real world" square wave requires some time, however small, to change from Value **A** to Value **B**. Rise time is the amount of time between the 10% and 90% points, as is fall time.

displayed on an oscilloscope screen is a true representation of what was in the circuit before he connected the oscilloscope to make the measurement? Some of the reservations the troubleshooter must keep in mind if he is to view what he sees constructively are listed below:

1. The so-called square wave is not square; rather, it has a certain amount of rise time and fall time, and the corners are rounded.
2. The oscilloscope used "colors" the waveform. Merely connecting the oscilloscope to make a measurement changes the circuit because the oscilloscope becomes part of the circuit. The oscilloscope also has its own limitations of frequency response, that is, how fast its display can rise and fall, built into its design.

3. Unless a compensated probe is used with the oscilloscope, the displayed waveform may be so greatly different from "what should be there" that it may be mistaken as a trouble symptom when actually it is not.

The TV troubleshooter today deals mostly with color receivers. For this reason a good oscilloscope is required, one that is able to show the color sync signal, or color burst, which is approximately 3.58 MHz in frequency (located on the "back porch" of the horizontal sync blanking pulse as shown in Figure 6-1c). In practical terms this usually means that the oscilloscope should have minimum frequency response capability in its vertical amplifiers to at least 5 MHz, with a compensated probe as a desirable accessory. Although monochrome TV servicing does not require such a wide-band oscilloscope, the troubleshooter always profits from being able to see high frequency signals better.

In any case, the troubleshooter should become familiar with the limitations of the oscilloscope he will be using. If possible, it is a good idea for him to make a side-by-side comparison with a really good oscilloscope to see how much his own machine colors the observed waveform and how much high frequency detail it misses. Most important, he should learn what his own oscilloscope display will look like when viewing actual, *good* sync signals, color bursts, square waves, and other pulses.

The compensated probe maximizes the inherent capabilities of an oscilloscope by minimizing phase shifts at the oscilloscope input. Figure 6-3 illustrates this point.

If you are a beginning TV troubleshooter you should look at horizontal sync pulses and the composite video signal at different points. See if you can capture the color bursts. Assess how much your oscilloscope colors the observed waveform and how much it misses of what is there. After you have done this you are in a much better position to successfully pursue horizontal sync problems.

6-3 Separation of Sync Pulses from Video Information

To understand more clearly what is required of the sync separator, let us examine what it must do:

1. The sync separator must accept at its input terminal the total composite video signal from the video detector or video amplifier and remove video information, permitting only sync pulses to appear at its output terminals.

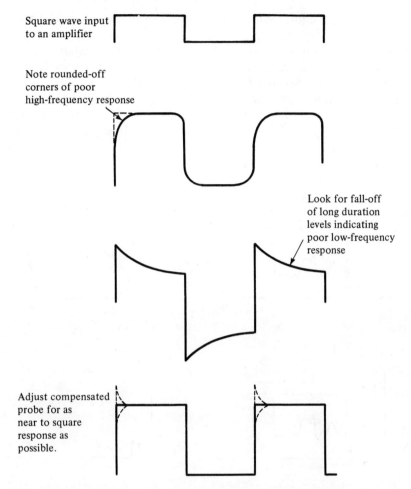

Figure 6-3 Pulses with short rise and fall times are a challenge to capture, to display accurately, and to interpret correctly because of limitations of the amplifier inside the oscilloscope.

2. The sync separator must faithfully preserve the timing integrity of each sync pulse and must not introduce any factor which varies it.
3. The sync separator must perform its function satisfactorily on all normal strength signals, ranging from weak to strong stations.

Sync separator circuits are designed to pass only those signals above a certain threshold voltage level which self-adjusts automatically to accommodate both strong and weak stations. This is accomplished by making the bias of the sync separator stage dependent upon the strength of the composite video signal entering the sync separator. Thus, on a strong input signal the bias is automatically larger, permitting only large signals to pass and cutting off small signals. On a weak signal, the bias is lowered, but still only enough to pass the sync pulses. Figure 6-4 illustrates the action in which the DC bias automatically adjusts to the proper level.

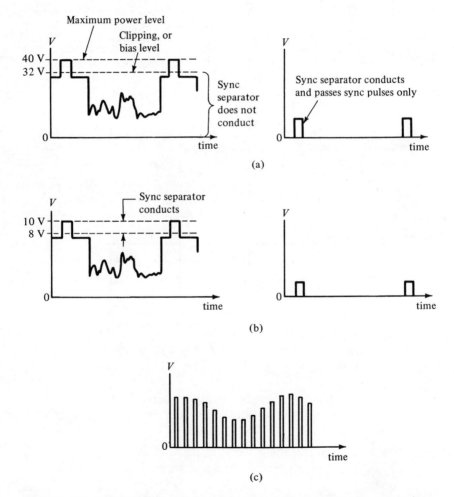

Figure 6-4 (a) Clipping or bias level self-adjusts to pass only sync pulses on strong signal. (b) Clipping level automatically adjusts for weak signal. (c) Pulse height with 60 Hz interference mixed with the sync.

6-3 Separation of Sync Pulses from Video Information 147

To view the horizontal sync pulse, the oscilloscope horizontal sweep frequency is adjusted to one-half the horizontal sweep frequency of the TV receiver, or 7875 Hz. The display is compared with "what should be there" according to service notes. If the pulse height is too small, the picture will roll easily and cannot be locked in solidly. When these symptoms are present the troubleshooter checks for proper pulse height first, then for the subsystem where the pulse height is lost. (It is possible for the horizontal AFC stage to be at fault if the pulse height is normal; this will be discussed in Chapter 8 where the horizontal sweep system is covered.) At the same time that the pulse height is checked, the operator also looks closely to see if all video information has been removed properly by the sync separator.

The majority of sync problems fall into two categories as follows:

1. *Partial or complete loss of sync.* This causes the picture to fall out of lock easily, or it may not lock at all. This can affect the vertical only, the horizontal only, or both, depending on where the fault is located (Table 5-1, 8 and Table 6-1).
2. *Problems involving timing.* The picture may lock in solidly, but picture elements may be displaced vertically or horizontally from different causes (Table 5-1, 9 and 10):
   a. *Ripple in the sync* from the AC power line or DC power supply (60 Hz or 120 Hz) can interfere with vertical timing or may cause one or two bends in left and right margins and in vertical lines in the picture (see Figure 6-5a and b).
   b. *Video in the sync* causes horizontal displacement of picture elements which changes as picture content changes. This can result from loss of bias on the sync separator, video amplifier, RF or IF amplifier, often through improper setting of the AGC control.
   c. Noise spikes in the signal.

Problems involving loss of sync are usually easy to isolate. The oscilloscope is used as a signal tracer to examine the composite video signal as it emerges from the video detector or the video amplifier and at the input of the sync separator; the separated sync is then traced to its termination points, the horizontal AFC circuit and the vertical oscillator circuit.

Sometimes the sync pulse amplitude may be compressed as a result of an amplifying stage that develops limited dynamic range or distortion. This will cause the picture to roll easily, depending

TABLE 6-1

Flow Chart for Troubleshooting Television Sync Problems

Problem: Neither vertical or horizontal hold controls can be adjusted to solidly lock in the picture, which appears normal otherwise.

Possible causes: Partial or complete loss of sync signal somewhere between take-off point on video amplifier and sync separator output due to defective sync separator tube, transistor, or a fault in the circuit.

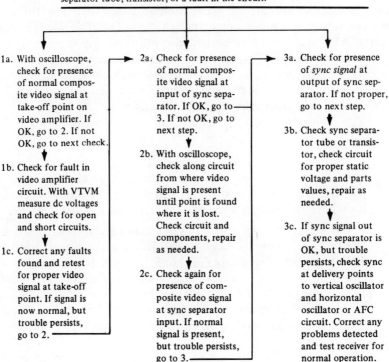

INSTRUMENTS:
Oscilloscope
Tube or transistor checker
VTVM

SERVICE NOTES:
Refer to service notes for circuits, signal levels, test points, dc voltages and component values.

on the severity of sync compression. This trouble is usually traceable to a defective tube or transistor, or to a changed value component that upsets the essential static conditions (DC voltages) somewhere within the circuit. Among components, leaky capacitors are the most frequent offenders.

6-3 Separation of Sync Pulses from Video Information

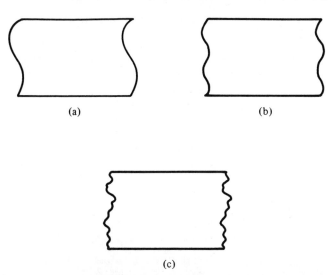

Figure 6-5 (a) Bending in raster due to 60 Hz from poor power supply filtering or AC from the power line varying the horizontal timing. (b) Bending in raster due to 120 Hz varying the horizontal timing, usually due to poor filtering of DC power supply. (c) "Tearing" in the picture due to video information getting into the sync. Look for trouble with clipping level (bias) in sync separator. Note that the bending in both (a) and (b) will drift upward through the picture on color transmissions.

When troubles involving timing (see Category 2 above) are present, the challenge is to identify the nature and location of interference that is disturbing the sync pulse. Figure 6-6a shows noise pulses that occur just prior to the vertical sync pulse. These can trigger the vertical oscillator prematurely and cause the picture to roll vertically. Noise can usually be identified as the trouble source by careful observation of the picture and the manner in which it rolls. Common noise sources are brush type electric motors, auto ignition systems, neon and fluorescent lights, and the AC power line. This type of noise interference enters the receiver via the antenna, and there is little one can do to correct the problem once the noise signal has entered the set.

If the offending motor or other noise producer can be located, it is best to try to filter out the noise at the source. Noise from the AC power line is usually beyond the scope of the TV troubleshooter to cope with successfully. If the noise is excessive, a complaint should be made to the power company asking them to investigate. This will sometimes bring corrective action.

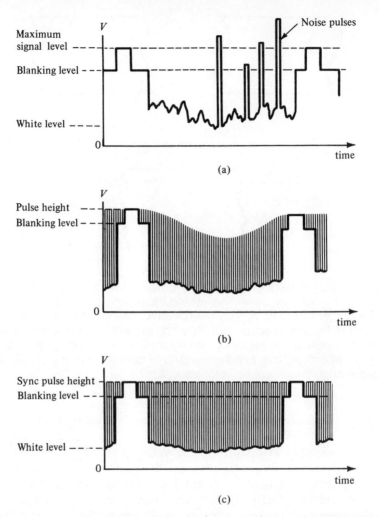

Figure 6-6 (a) Noise pulses in signal which interfere with the information on the carrier. (b) Interfering AC voltage causing a "hill and dale" effect seen when viewing the vertical sync pulse. (c) Normal composite video signal. Note that pulse height is steady and unvarying.

Variation in voltage or pulse height at the input to the sync separator tends to appear as variation in height of the sync pulse following the sync separator. This later variation in turn usually translates as variations of the precise triggering times of the sweep oscillators.

Loss of filtering or inadequate filtering in the DC power sup-

6-3 Separation of Sync Pulses from Video Information

ply produces a 60 Hz ripple in half-wave rectifiers and 120 Hz ripple in full-wave types. This ripple appears as variations in the B+ voltage (see Chapter 1 for methods of checking) and influences amplifier gain. The net result is typically a "hill and dale" effect noted in the composite video signal at the output of the video detector or video amplifier, which is seen at the input to the sync separator. This is shown in Figure 6-6b. This type of interference with timing usually shows up as a single bend at the left-hand margin of the raster (60 Hz) or two bends (120 Hz). On color signals these bends do not stay stationary but drift slowly upward when the picture is locked in; this is the clue that tells the troubleshooter that the nature of the interfering voltage is connected with the AC power line or filtering of the DC power supply.

If two bends in the picture are noted, check the power supply. It is the logical trouble site as it is the most likely, or only, source of 120 Hz ripple when it is a full-wave rectifier. If one bend is noted, check to see if half-wave rectification is the cause. If ripple on the B+ agrees with service notes on what should be there, look for leakage from the power line itself. In vacuum tubes, heater-cathode leakage or shorts are the most likely cause.

When video becomes mixed with the sync, the sync voltage will vary with the picture content, causing the timing to vary constantly. Video in the sync is usually diagnosed by observing the television picture, because the trouble typically causes the picture to tear horizontally to the right as the picture changes or as figures move in the picture. Because video information is fleeting and constantly changing, it is more difficult to capture and identify using the oscilloscope than by observing the symptoms produced in the TV picture itself.

A final comment is appropriate at this point concerning sync signal processing *after* the sync separator stage. The separated horizontal sync pulse is passed through a differentiator network, as illustrated in Figure 6-7a, before it reaches its termination point in the AFC circuit. The differentiator accepts the square-wave sync pulse at its input and produces a spike at its output. It is, in fact, a high-pass filter. On the other hand, the vertical sync pulse passes through an integrator circuit following the sync separator before it triggers the vertical oscillator. The function of the integrator circuit is to "sum up" the average voltage of the series of pulses that constitute the vertical sync pulse in its entirety. Refer to Figure 6-7b for details. The specific characteristics of the vertical sync pulse are discussed in fuller detail in the next section.

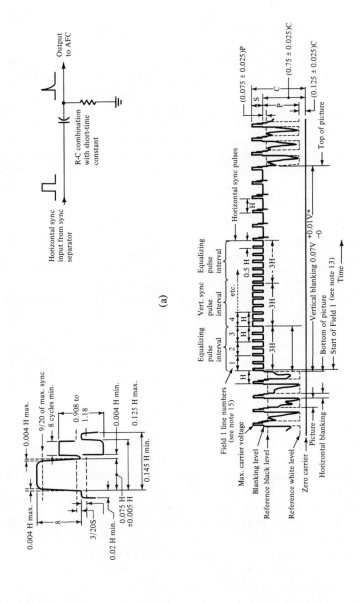

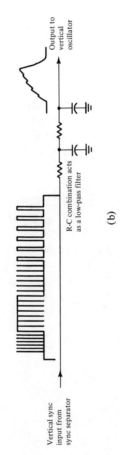

Figure 6-7 (a) Horizontal sync pulse and differentiator network. (**Note:** H = 1 horizontal period.) (b) Vertical sync pulse and integrator network. (**Note:** V = 1 vertical period.)

6-4 The Vertical Sync Pulse

The vertical sync pulse is more complicated than the horizontal sync pulse due to a number of factors. The most important of these is that horizontal sync and timing must be maintained throughout the duration of the vertical sync and blanking pulse, which is much lower in frequency and longer in duration than the horizontal. Both vertical and horizontal oscillators operate continuously while the receiver is operating. To see how these problems are resolved, let us re-examine the way the picture is made and the timing details, presented in Figure 6-7.

We recall that 30 pictures are transmitted per second, and each picture is scanned twice, making the vertical scanning frequency 60 Hz. We also remember that we allocate 525 lines per picture, to be divided between the two scanning fields. This means that each vertical scanning field will consist of 262.5 horizontal scanning lines. The system calls for *interlaced scanning* to minimize flicker. In interlaced scanning the horizontal lines of the odd-numbered vertical scanning fields are timed so that they fall midway between the already-produced lines of the previous even-numbered fields which are still visible. It is the job of the short duration horizontal sync pulses and equalizing pulses that are transmitted *during the vertical blanking period* to accomplish the precise horizontal timing that is necessary to make interlaced scanning work.

The troubleshooter is not much concerned with these details, other than to understand how the system works. But he is vitally concerned that the functions that these details are a part of are working as intended. This is the reason his primary attention is focused so heavily upon trouble symptoms and check points, because they tell him whether the functions are being carried out properly. If functions are not being correctly fulfilled, he then looks to details of circuitry to find out what went wrong.

The vertical integrator network lies between the sync separator and the vertical oscillator. Its function is to act as a low-pass filter, removing high frequency (short duration) pulses such as the horizontal sync and equalizing pulses, as well as most noise. Only the vertical sync pulse with its greater period and lower frequency should pass successfully through the integrator. Even so, the pulse delivered to the vertical oscillator is not large, typically in the range of 0.4 to 4.0 volts peak to peak.

It is easy to observe the vertical sync pulse where it enters the vertical oscillator if the oscillator is *not* operating, but difficult if the oscillator *is* operating. This is because the vertical oscillator produces a much larger pulse that masks the sync pulse. The result is that the troubleshooter cannot easily examine the sync pulse while the oscillator is running. So, the vertical oscillator must be disabled to make this particular check. Service notes give the recommended way to disable the vertical oscillator for a particular TV receiver. However, in the absence of such instructions it is usually safe and effective to kill the vertical oscillator in tube type receivers by bypassing the signal entering the vertical amplifier stage by connecting a large capacitor from the grid to ground. Refer to Figure 6-8 for details. In transistorized receivers, it may not be necessary to disable the vertical oscillator to see the sync pulse; refer to service notes for instructions.

6-5 Automatic Gain Control in Television Receivers

The purpose of automatic gain control, or AGC, is to limit the variation in signal strength presented to the video detector to a narrow range for both strong and weak stations. Effective AGC will produce approximately the same picture contrast for both strong and weak stations without readjusting the controls.

In older tube type receivers, a negative DC voltage in the range of -0.4 to -4.0 volts is used to control the gain of the RF amplifier and one or more IF amplifier stages by varying their bias. This voltage is minimum on weak signals, permitting maximum gain, and larger on strong signals, which reduces gain. The voltage adjusts automatically as the antenna signal varies in strength, which gives it the designation automatic gain control.

The AGC voltage may be developed in different ways. One early method was similar to automatic volume control (AVC) in radio; that is, the output of the video detector was filtered of its video and sync signal variations and then applied to control gain. The major disadvantage of this method was that it was slow acting, which permitted "airplane flutter" in the picture to become very pronounced. Later circuits employ "keyed" AGC which is much faster acting and produces a much more stable picture. In both tube and transistor circuits, keyed AGC requires two properly timed voltage pulses to be applied to a keying tube or transistor in order to work. One of these is a constant voltage level pulse that comes from the horizontal sweep system, often from a winding on the horizontal output transformer. This pulse is usually ap-

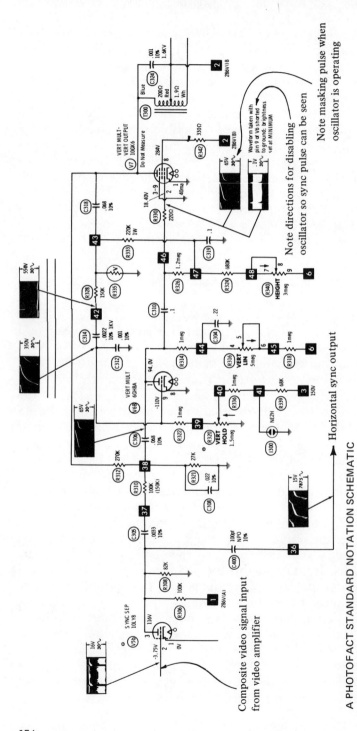

Figure 6-8 Vertical oscillator must be disabled to remove masking pulse in order to see sync pulse (courtesy of Howard W. Sams & Co., Inc.).

156

plied to the plate of the keying tube or to the collector or emitter of a keying transistor. The second pulse voltage is the horizontal sync voltage contained in the composite video signal. This is applied to the control grid of the keyed tube and to the base of the keyed transistor.

The pulse from the horizontal sweep system and the pulse from the transmitter-sent signal must be synchronized in order for the keyed tube or transistor to conduct. That is, there must be positive-going plate voltage *at the same time* that there is positive-going grid voltage from the composite video signal; otherwise, there will be no plate current and no AGC voltage. Assuming that both pulses are timed right, or keyed, the amount of current that will flow depends on the magnitude of the voltage of both sync pulses and the setting of the AGC control by the technician. See Figure 6-9 for details.

Summing up, AGC voltage will be produced in a keyed AGC system only when the following conditions are met:

1. *A pulse from the horizontal sweep system must be present and a station must be tuned in to provide sync pulses.* Absence of either pulse means that no AGC voltage will be produced.
2. *Both pulses must occur at the same time.* If the horizontal oscillator is on the wrong frequency and the picture is not locked in horizontally, no AGC voltage will be produced.
3. *The AGC control must be properly adjusted.* Too much AGC voltage will create insufficient contrast, snow in the picture, or no picture at all; too little AGC voltage will produce excessive contrast, usually sync buzz in the sound, often accompanied by horizontal tearing and instability of the picture.

6-6 Adjusting and Troubleshooting AGC Circuits

When a television receiver is first installed the AGC control is set at the optimum level by the following procedure:

1. Tune the set through all the stations on the air in the locality and determine which signal is strongest.
2. Set contrast control to maximum, then adjust the AGC con-

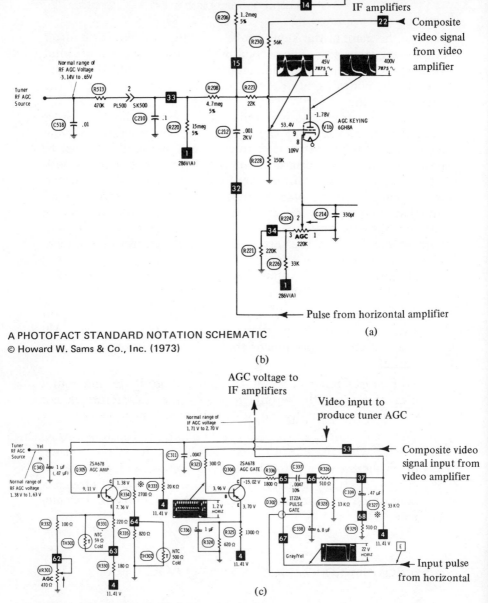

Figure 6-9 (a) Two pulses are required in keyed AGC. (b) One version of keyed AGC circuit in which a vacuum tube is used. (c) Keyed AGC in a transistor receiver. **Note:** Only the IF AGC voltage is produced by the keyed circuit, and tuner AGC is not keyed (courtesy of Howard W. Sams & Co., Inc.).

6-6 Adjusting and Troubleshooting AGC Circuits 159

trol until excessive contrast, sync buzz, and sometimes picture tearing is noted. Next, readjust the AGC control until all sync buzz and tearing disappears and the contrast is high, but not excessive.

3. Tune through all stations once more, watching and listening carefully for any sign of sync buzz, tearing, or excessive contrast on any station. If any stations show these symptoms, readjust the AGC control carefully until they disappear, but no further. This is the optimum setting for the AGC control.

When trouble in the AGC circuit is suspected, it is a good idea to begin troubleshooting by turning the AGC control through its range while watching the picture closely to see if the trouble clears up or persists. If the trouble persists, the next step is to check the keying tube (if tube type) or the composite video signal (if transistorized) at the output of the video detector. If the oscilloscope shows the composite video signal to be normal the trouble is not in the AGC circuit. If the composite video signal is not normal, the trouble is in the tuner, the IF amplifiers, or the video detector and may, or may not, be an AGC problem. (Refer to the Troubleshooting Flow Chart of Table 6-2 for procedures and suggestions.)

After trouble has been confirmed to exist prior to the video amplifier, the question to be resolved is whether the trouble is AGC-connected or not. One method of approach is to use a "bias box" to substitute a small, controllable DC voltage in place of the AGC voltage normally produced by the receiver itself. The bias box is a small, adjustable low voltage DC power supply manufactured expressly to supply bias voltage for testing and alignment. The bias box is connected to the AGC buss line, and then its output voltage is adjusted up and down. If the TV picture clears up and becomes normal, one concludes that something has gone wrong in the receiver's built-in AGC circuit; if the picture does not clear up, the troubleshooter looks for circuit trouble that is not AGC connected, following standard troubleshooting procedures of checking dynamic and static conditions.

Some AGC circuits connect to voltage sources of different levels and at several points. These may be difficult to deal with, especially if the trouble is a cumulative result of small changes in voltages and component values, and no single one by itself is significant. Patience, careful consideration of the data, and persistence will turn these frustrating cases into success stories. Remember, they *did* work once!

TABLE 6-2

Flow Chart for Troubleshooting AGC Problems in TV

Problems: Picture contrast is excessive, and picture tears intermittently.
Possible causes:
1. AGC control is improperly adjusted.
2. Defective AGC keying tube or transistor.
3. Trouble in AGC circuit; there is a circuit or component breakdown.

1a. Turn AGC control through its full range. If picture becomes normal at some point, you have probably found the trouble; go through set-up procedure for optimizing AGC control setting. If trouble persists, move to 2.

2a. Check vacuum tubes by testing or substitution in the following circuits: AGC, RF stage in tuner, 1st and 2nd IF amplifiers. Replace all that are defective. If trouble persists, go to next step.

2b. With oscilloscope, check to see if both the horizontal pulse from the sweep system and composite video signal with sync pulse are reaching the keyed tube or transistor. If not, the cause must be found and corrected. If both pulses are OK, proceed to next step.

2c. Connect bias box to AGC line at output point of AGC filter. Vary the voltage applied to the AGC line, noting if the picture at some point becomes normal. If it does, trouble is likely to be in the AGC circuit somewhere. Go to 3. If trouble persists, the fault is most likely to be found in the tuner or IF amplifiers. Follow standard troubleshooting procedures to pinpoint the defect.

3a. Measure the AGC voltage with VTVM at various points along the AGC line as you vary the bias box. Note conditions when picture is good. Remove bias box and go to next step.

3b. With receiver's own AGC operating check AGC voltages and compare with those present when bias box produced a good picture and with values given in service notes. Analyze cause of any discrepancies, check and repair as needed.

INSTRUMENTS:
Oscilloscope
VTVM
Tube checker
Transistor checker
Bias box

chapter 7

The vertical sweep circuit

The function of the sweep circuits in television is foreign to radio and audio equipment. While it is fairly simple to describe the role of the sweep circuits in concept, their function is not always so easy to achieve and maintain in practice. Briefly, vertical and horizontal sweep circuits have certain requirements in common that they must meet:

1. The sweep circuits must move the lighted spot at a uniform rate of speed across the face of the picture tube. Failure to do this will result in nonlinearity and distortion of figures and other picture elements.
2. The sweep circuits must complete their retrace in the allotted time. Failure to do this may cause the retrace to be visible because the blanking pulse has ended; this usually appears as a sort of lighted film at the left side of the raster and picture if it is the horizontal retrace that exceeds the blanking time. (Vertical retrace is blanked out by a built-in circuit that is not dependent on the blanking pulse in modern receivers.)
3. Any "ringing" or oscillation induced by the rapid changing of magnetic and electric fields during retrace must be killed off by the end of the blanking pulse. Failure to do this will result in interference with the picture.

Because the collapse of the magnetic fields is so rapid during retrace in both vertical and horizontal sweep circuits, very high pulse voltages are induced. These high voltages place much stress on circuits, components and insulation, which makes the sweep circuits the most breakdown-prone subsystems in the TV receiver. In fact, these voltages are so high that they pose a very real potential for damage to measuring instruments. Only in TV do we find in service notes the admonition, *"Do Not Measure"!*

7-1 The Nature of the Vertical Sweep Circuit

The vertical oscillator is a nonsymmetrical multivibrator. This means that two vacuum tubes are normally used if the receiver is tube type, while two or more transistors are used in solid state sets. "Nonsymmetrical" merely means that the "on time" of the output tube or transistor represents the *trace* time and is longer than the "off time", which represents the retrace time. Remember that the trace time represents the drift of the beam downward through the picture, whereas the *retrace* time represents the rapid return of the beam from the bottom to the top of the picture. When the output tube is "on", the other tube is "off", and vice versa. Figure 7-1 presents two representative vertical sweep circuits, one tube type and the other transistorized. These are from the same receivers used in the previous chapter on sync and AGC.

The duration of "on-time" and "off-time" in both vacuum tube and transistor circuits is determined by R-C time constants, that is, by the values of resistors and capacitors that make up the circuit. Thus, any change in either resistance or capacitance changes the period and frequency of the multivibrator and causes the picture to roll vertically. The familiar vertical hold control, in fact, is a variable resistor which is adjusted to establish the proper R-C time constant required to produce the desired vertical frequency, which locks in the picture.

Even with the vertical hold set properly, the vertical sync pulse is still required to hold the receiver scanning precisely in step with camera scanning at the transmitter. Since the free-running frequency of the oscillator is set slightly lower than the required sweep frequency, a function of the sync pulse is to "nudge" the multivibrator into conduction slightly sooner than it would otherwise conduct while free running.

The output of the vertical oscillator is transformer coupled to the vertical yoke coils in tube type receivers. In solid state receivers, it may be capacitively coupled. The coils themselves are part of the deflection yoke, which they share with the horizontal coils. The coils are located in the circuit, as shown in Figure 7-2. Rising current in the coils creates an increasingly strong magnetic field. This permeates the neck of the picture tube immediately under the yoke. The electron beam passes through this region, its path deflected in proportion to the strength of the magnetic field. The stronger the magnetic field becomes, the more the beam is bent. In this way, both vertical and horizontal deflection are accomplished.

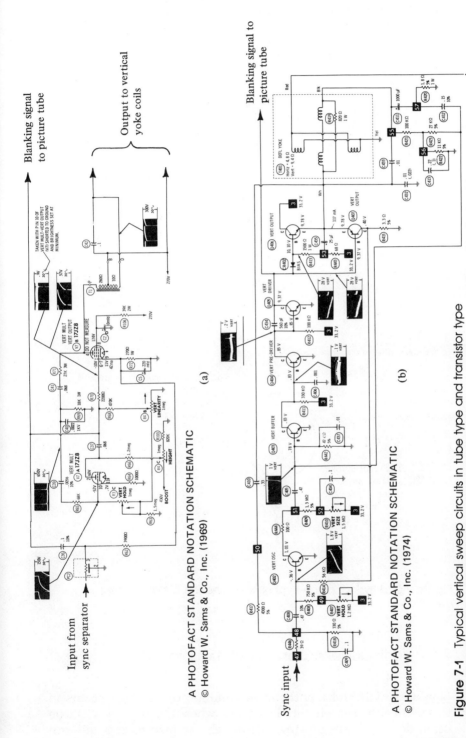

Figure 7-1 Typical vertical sweep circuits in tube type and transistor type TV receivers (courtesy of Howard W. Sams & Co., Inc.).

164 The Vertical Sweep Circuit

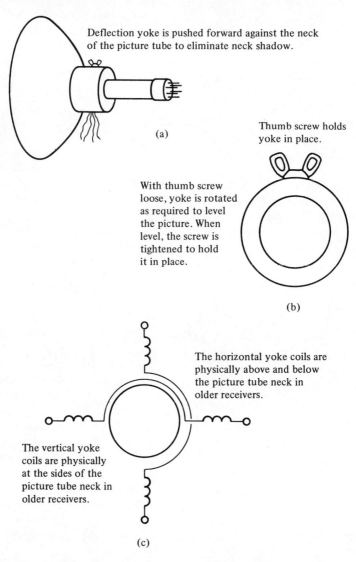

Figure 7-2 Physical placement of yoke and the deflection coils.

7-2 Vertical Deflection Circuit Troubles

Nonlinearity is probably the most common trouble that occurs in the vertical sweep circuit. It results when the moving electron beam does not drift downward through the picture at a uniform

7-2 Vertical Deflection Circuit Troubles

rate during trace time. Distortion in the picture is seen because the horizontal scanning lines are not spaced evenly apart from top to bottom of the raster as they should be.

The most frequent causes of vertical nonlinearity are changes in the characteristics of vacuum tubes and changes in the values of resistors and particularly of capacitors. Transistors tend to be either good or bad, although they are subject to in-between changes from excessive heat. Occasionally, part or all of a yoke coil will short, changing the magnetic field and with it the deflection and the raster.

The time versus magnitude characteristic of the driving signal to the vertical output amplifier that is displayed by an oscilloscope is very helpful in analyzing nonlinearity. Note the waveform at the signal grid of the vacuum tube in Figure 7-1a and at the base of the transistor in Figure 7-1b; both show a short period of retrace time and a long, gradual change in voltage for the trace time.

If the rise in voltage and current varies from that which is specified, the deflection will be affected. Figure 7-3a illustrates the effect on a circle if the driving waveform should rise rapidly at the beginning of its downward scan (which begins at the top of the picture) then slow to normal for the last part of the scan. Figure 7-3b shows the effect on a circle if the scanning is normal at first and then its rate of change increases for the final portion of the scan (which is the bottom of the picture). Figure 7-3c shows the effect that a slower than normal rate of change would have if it should occur in the middle of the trace time. Note that the foregoing examples are exaggerated to better illustrate the principles involved. It is imperative that the rate of change of the magnetic field deflects the scanning beam at a uniform (linear) rate. If this is not done, distortion of the picture is a certain result.

Reasoning backward from symptom to cause, when vertical nonlinearity is seen in the picture the troubleshooter concludes that something has affected the rate of change of the yoke current, or at least of the magnetic field, and he searches for the specific cause. For suggestions relating to checks and sequences to follow in troubleshooting problems involving vertical nonlinearity, refer to Table 7-1.

Troubles involving frequency become critical when the picture can no longer be locked in by adjusting the vertical hold control. When this symptom is seen, it always means that something has happened to change the time constants in the vertical multivibrator circuit. This can mean a change in either a resistor (or resis-

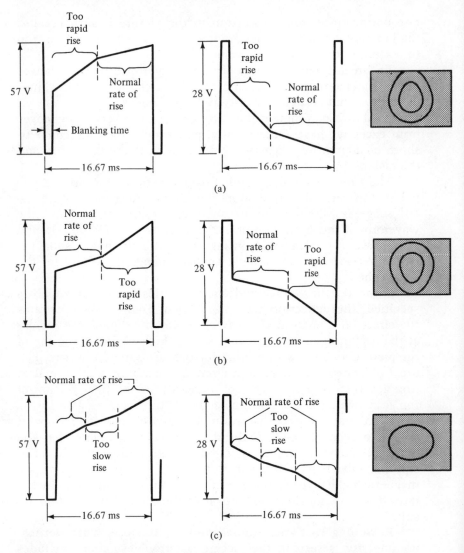

Figure 7-3 Graphic illustration of the effects on the picture of a circle if the rate of change of voltage and current in the deflection yoke (vertical) is not linear. (**Note:** Voltage levels are taken from Figure 7-1. Waveforms have been modified to illustrate the topic under discussion.)

tors) or a capacitor (or capacitors), or both. The troubleshooter's problem is to discover which one(s) have changed value.

Capacitors break down more frequently than resistors in the vertical circuit. To see why, first locate the plate of the vertical output tube in Figure 7-1a. Note the caution, "Do Not Measure,"

TABLE 7-1

Flow Chart for Troubleshooting Vertical Sweep Circuit Linearity Problems

Problem: The vertical linearity is poor.
Possible causes:
1. Vertical linearity and vertical height controls are improperly adjusted.
2. A vacuum tube is defective.
3. A capacitor has become leaky or a resistor has changed value.
4. There is a short or open circuit in the deflection yoke.

Check vertical height and vertical linearity adjustments to see if they correct the problem.
Check vacuum tubes next, preferably by substitution. Replace any that are questionable and test run the receiver.
With VTVM, check dc voltages at tube or transistor terminals and compare with values specified in service notes. Check for cause of any discrepancies greater than 15% and replace any defective components found.
Check for leaky coupling capacitors and for leaky capacitors in the feedback loop. Check resistor values. Replace any defective components found, then test for proper operation.
With the oscilloscope, check the driving signal to the vertical output stage. If OK, check the deflection yoke with an ohmmeter, or by injecting a signal from a yoke tester or by substituting a new yoke as a conclusive test.
If nonlinearity still persists, try substituting different values of resistance and capacitance by the use of decade boxes. Sometimes the cumulative effects of many small component changes create a problem which cannot be traced to a single part alone.

at this point. Note, also, that the feedback to enable the multivibrator to operate comes from this same point and goes through R71 and C41 to a junction of R69, C40, and C39. C39 in series with R65 completes the feedback path to the grid of the vertical multiplier tube, although R3C, the vertical hold control, and R4B, the vertical linearity adjustment, along with R63 complete the circuit to ground. We mention all these components to emphasize that the "Do Not Measure" pulse is applied to this entire net-

168   The Vertical Sweep Circuit

work! Is it surprising that one of these capacitors should sometimes short or become leaky?

For troubleshooting suggestions and sequences involving a vertical frequency problem, refer to Table 7-2. This is a flow chart that represents one approach to solving vertical sweep circuit frequency problems with the minimum number of instruments, just an oscilloscope and a VTVM.

7-3 Special Procedures, Hints, and Difficulties

First of all, whether the problem is with vertical linearity or with vertical frequency, the troubleshooter should check all vertical controls and run each control through its entire range — the vertical hold, the vertical height, and the vertical linearity (if present) — and note any effects on the picture that might constitute a

TABLE 7-2

Flow Chart for Troubleshooting Vertical Frequency Problems

Problem: Vertical hold control will not stop picture from rolling.
Possible causes: 1. A capacitor or resistor has become defective.
2. A number of circuit components have changed slightly in value due to aging, and the problem is the cumulative result of these small changes.

Check dc voltages at tube or transistor terminals and compare measured values with voltages specified in service notes. Investigate the cause of any significant discrepancies found and repair as necessary. Test for proper operation.

↓

Check carefully for leaky capacitors and for changed value resistors. Replace leaky capacitors and resistors whose values are out of tolerance. Then recheck the receiver for proper operation.

↓

If trouble persists, use capacitance decade to substitute for frequency determining capacitors in the circuit. Vary the capacitance until vertical frequency is correct and can be controlled by the vertical hold control. When the best value of capacitance is found, a capacitor of that value should be permanently installed and then the receiver rechecked for proper operation.

↓

If the last step failed to correct the frequency problem or resulted in nonlinearity, it may be necessary to substitute resistor values as well as capacitors. Be assured that there is a proper RC combination that will correct the trouble.

7-3 Special Procedures, Hints, and Difficulties 169

clue to exactly where the fault lies. The vertical height and vertical linearity controls are both variable resistors, and often have a pronounced effect on the frequency as well as the picture linearity. In such a case it may be necessary to adjust all three controls experimentally, a little at a time, to reach the proper settings that will result in a good picture.

Now, a comment on electrolytic capacitors, such as the cathode by-pass capacitor (C5) and emitter by-pass capacitor (C435). Electrolytic capacitors (Figure 7-1a and b) tend to dry out and decrease in size with age, which impairs their ability to bypass variations in voltage. Usually this will decrease the vertical sweep magnitude, causing a black line at the bottom or top (or both) of the picture. Sometimes a defective electrolytic capacitor also causes nonlinearity.

When we check a by-pass capacitor in the circuit, we recall that a capacitor is supposed to act as an open circuit for DC and almost a short for AC. This means that if it is good (open) for DC, the same differences in DC voltage should be measured across it as service notes call for; if it is shorted the measured difference will be less. If the capacitor has not diminished in size, approximately the same variational (AC) voltage should be seen on both ends of the capacitor when viewed via the oscilloscope. If a difference in voltage is noted on one end compared to the other end, substitute a new one in parallel and recheck.

Problems that are heat-induced typically produce symptoms in which the picture changes as time passes and heat builds up. That is, the picture refuses to stay in proper adjustment for any length of time. Heat-induced problems may affect the linearity, the height and/or the vertical frequency. Such problems arise when the values of resistors, capacitors, and amplifying devices are affected by the temperature around them. To attack heat-induced problems, check vacuum tubes first, by substitution if possible. If the problem persists, indicating the tube was not the cause, next check capacitors and resistors. Often, changes in component values that noticeably affect the picture are not large enough to produce significant changes in static conditions that will enable the troubleshooter to identify the faulty part wth certainty.

One very effective method of identifying changed value components due to heat is to use a refrigerant which is manufactured in small spray cans expressly for this purpose. The refrigerant is sprayed on the resistor or capacitor suspected of changing value with heat, and the picture is watched carefully as the refrigerant

170 The Vertical Sweep Circuit

takes effect. If the picture changes, replace the component and recheck. Do *not* spray refrigerant on hot vacuum tubes as it will cause the glass to crack and destroy the tube.

If the symptom is a single bright horizontal line showing across the middle of the picture tube, it means that the vertical sweep has failed completely. The cause is probably catastrophic failure of some part or parts of the circuit. In most cases, a check of static conditions, DC voltages first, will reveal some incorrect values compared to what should be there. These are analyzed for possible causes, then the possible causes are checked out, one by one, until the fault is found. A clear-cut problem such as no vertical sweep is generally among the easier problems to cope with successfully.

All of our discussion so far has been based upon *signal tracing*. This is because we are dealing with an oscillator, a circuit which generates its own signal. *Signal injection* as a technique has not been mentioned, but do not take this as an indication that signal injection has no place in troubleshooting vertical sweep circuits. Indeed, signal injection can be very useful in determining where in the vertical circuit the trouble exists. However, it does require a special type of signal generator that can produce the same waveforms as the vertical circuit normally produces. Several companies manufacture signal generators for this application.

The advantage of using signal injection in the vertical sweep circuit lies in its ability to pinpoint troubles in the yoke, the output transformer, and the output stage. The key to its effectiveness is that a known, good signal from the signal generator is *substituted* for the signal produced by the receiver's own vertical oscillator and sweep circuit. Thus, the signal can be fed to the yoke, to the output transformer, or into the vertical amplifier as its driving signal, and the raster and picture checked for normalcy. The disadvantage of signal injection is that it requires another piece of specialized test equipment whose expense may be difficult to justify unless one is doing a significant volume of TV repair work.

Without signal injection, we must depend on static checks of voltages, parts values, resistance and circuits, aided by dynamic checks using the oscilloscope. Data obtained in this fashion is not always a clear-cut indicator of the defective component because different causes can sometimes yield the same data, so in these cases component substitution has to provide the definite answer.

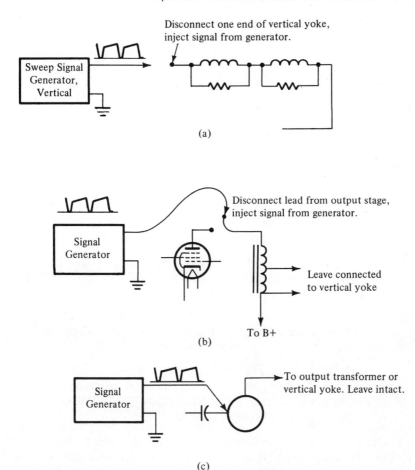

Figure 7-4 Signal injection technique to pinpoint source of trouble. (a) Checking the deflection yoke. (b) Checking the output transformer and deflection yoke. (c) Checking the amplifier, transformer, and yoke. (**Note:** Always check applications notes of generator before using for the first time to be certain that it is being used and connected correctly.)

Frequently, several components may have to be tried before the defective one is found.

Two very useful pieces of equipment when working with vertical deflection circuits are a capacitor decade box and a resistance decade box. Their great advantage is that they can be substituted for a particular capacitor or resistor in the circuit and so a range of values can be tested under working conditions by merely flicking a switch. Where the problem is the result of cumulative minor changes in several circuit values, this offers a rapid way to experimentally determine what values of resistance and capacitance will restore the circuit to proper operation.

7-4 Summary

The vertical sweep circuit is a nonsymmetrical multivibrator. Its frequency depends upon R-C time constants in its circuit, and its linearity depends upon the yoke producing a properly varying magnetic field to deflect the electron beam downward across the picture tube at a uniform rate.

The most common troubles in vertical sweep circuits fall in these categories:

1. *Nonlinearity—picture proportions are distorted.* Refer to Table 7-1 for troubleshooting suggestions.
2. *The picture rolls and cannot be stopped, i.e., the vertical frequency is wrong.* Look in Table 7-2 for suggestions and sequences.
3. *No vertical sweep.* This is the easiest trouble to correct. Use standard techniques of checking static conditions, circuits, and component values. Refer to Section 7-3 and, if necessary, to earlier chapters on static checks.

Capacitors tend to become shorted or leaky due to high pulse voltages in many vertical circuits. This may affect frequency, linearity, or both, depending on which capacitor fails. Resistors sometimes develop a tendency to change value with heat, as do other components. Spray-on refrigerant can help to identify the cause of heat associated problems.

Signal tracing via the oscilloscope and checks of DC voltages, component values, and circuit continuity using the VTVM or a similar instrument are the backbone of vertical circuit troubleshooting. Signal injection (substitution) is useful, although it requires a special signal generator to produce the vertical waveforms.

chapter 8

The horizontal sweep circuit

The basic function of the horizontal sweep circuit is to move the lighted spot horizontally across the face of the picture tube at a uniform rate. In doing this the receiver, like the transmitter, must be synchronized and conform to established signal standards. But this is not all. The output of the horizontal amplifier provides the input both to the high voltage rectifier and to the sweep. In addition to that, the horizontal sweep feeds the damper circuit whose major purpose is to kill off ringing by the time the horizontal blanking pulse ends and whose minor purpose is to salvage some of the energy released by the collapsing magnetic field during retrace and apply it to some useful purpose. Yet another important part of the horizontal sweep system is the automatic frequency control circuit (AFC), which stabilizes the frequency of the horizontal oscillator. Figure 8-1 presents a functional block diagram of the complete horizontal sweep system.

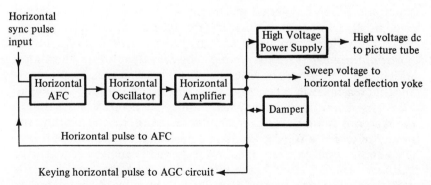

Figure 8-1 Functional block diagram of typical horizontal sweep circuit. Arrows indicate direction of signal energy flow and which circuit drives another.

174   The Horizontal Sweep Circuit

Considered in its entirety, the horizontal sweep system is probably the most complex circuit in the monochrome TV receiver. Because it includes the high voltage power supply, parts of the horizontal sweep system are subjected to the highest voltages found in the receiver. Troubles in this system are comparatively quite common, as one might expect. Fortunately, the most frequent troubles produce distinctive symptoms which makes the causes easy to locate and repair. Some of the less common troubles are more challenging.

In this chapter we will look at the horizontal sweep system, one function at a time, and point out the most common troubles and suggest ways to locate and repair the cause of each of them.

8-1   Troubleshooting the Horizontal Sweep System: Symptoms, Troubles, and Checks

The most common troubles in the horizontal sweep system produce symptoms that are quite distinctive and cannot easily be ignored:

1. No brightness, i.e., complete loss of raster.
2. Picture does not fill screen horizontally, but leaves black line at right or left margin, or both.
3. Focus is poor and usually changes with the brightness setting. The picture may get larger, or "bloom," and lose focus as the brightness is advanced, resulting in a loss of fine detail in the picture.

There are other problems, of course, such as poor linearity, but they are far less common than the troubles listed above.

If we reflect a moment, we can see that *all troubles in the horizontal sweep system produce visible symptoms.* Even arcs, which can sometimes be heard, also produce picture interference. For this reason it is especially important for the troubleshooter to look very carefully and thoughtfully at what he sees pertaining to the raster. A main point to remember, then, is that a malfunction in any one of the functional stages shown in Figure 8-1 will produce some kind of effect on the raster that can be seen. And, a failure of the horizontal oscillator, the horizontal amplifier, the damper, or the high voltage rectifier will produce a complete loss of raster, because all of these result in loss of high voltage.

The first checks that a troubleshooter makes are to localize the problem as rapidly as possible. Table 8-1 suggests some helpful steps and procedures for analyzing a common horizontal circuit trouble.

8-2 Picking a Starting Point: The Role of the Educated Guess

Whenever we have a problem before us, as soon as we have verified and defined its nature our thought processes take over to review what we know and what we need to know. For example, let us consider the problem of no brightness (loss of raster) and follow through the steps of Table 8-1.

First we consider the possible causes of what we see, of the visible symptoms. No brightness can result from lack of high voltage, wrong voltages at the picture tube socket, or a bad picture tube. We also know that picture tubes last for several years on the average, and that high voltage stresses are more prone to cause breakdown than lower voltages. So we decide to begin our checks with the high voltage system rather than the picture tube or the low voltages on the picture tube socket terminal. (We could have elected to start checks with either one of them; we are merely playing the percentages, making an educated guess.)

Since we have decided to start our checks with the high voltage system the next question is where. In TV, the most often used starting point is a check of the input to the high voltage rectifier anode. This is checked for "fire" as previously described in Chapter 5, Section 5-3, and Figure 5-2. This check helps to localize the problem. A lack of voltage here indicates trouble in the horizontal oscillator or amplifier, or possibly the damper; adequate voltage here means the horizontal oscillator, amplifier, and damper are functioning and the problem is in the high voltage power supply circuit, the low voltage circuits supplying the picture tube socket terminals, or the picture tube itself.

8-3 Checking the High Voltage Power Supply

The high voltage power supply owes its existence to the picture tube. That is, the picture tube will not light up until a source of high voltage DC is connected to its second anode. The amount of voltage required depends upon the size of the picture tube. Typi-

TABLE 8-1

Flow Chart for Troubleshooting the Horizontal Sweep System

Problem: No brightness, i.e., complete loss of raster

Possible causes:
1. High voltage power supply defective.
2. Horizontal amplifier circuit defective.
3. Damper circuit defective.
4. Horizontal oscillator circuit defective.
5. Picture tube defective.
6. Low voltage supply to picture tube base defective.

INSTRUMENTS:
VTVM,
oscilloscope,
tube or transistor checker,
flyback transformer and yoke tester,
picture tube checker.

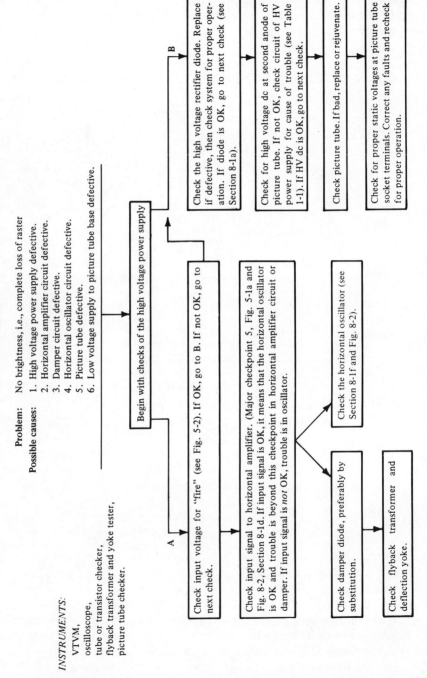

176

8-3 Checking the High Voltage Power Supply

cal values range from approximately 12,000 volts for small 12-inch tubes in portable receivers to 18 to 20,000 volts for 21-inch tubes. Larger tubes and color tubes require more, with many requiring DC in the 22,000 to 24,500 volt range.

The circuit of the high voltage power supply is a simple half-wave rectifier type in most TV receivers. The input to the rectifier diode comes from the horizontal output transformer when it is shocked into producing high voltage pulses during ringing. Ringing occurs as a result of the sudden collapse of the horizontal current and its magnetic field when retrace is initiated. Because the frequency of the high voltage pulses is so high (approximately 70,000 Hz), very small amounts of filter capacitance are needed. This is an advantage. In fact, the filtering is done by the capacitance between the inside and the outside of the picture tube bell (the glass is the dielectric) in many receivers.

Sometimes a resistor is found in series with the high voltage lead to the picture tube. Some receivers may have a separate filter capacitor in addition to the capacitance of the picture tube to provide high voltage filtering. In recent years, variations in the traditional high voltage supply have appeared. Some voltage multiplier circuits may be encountered, and solid state diodes are replacing vacuum tubes for both damping and high voltage rectification.

The high voltage supply, like any other supply, must have proper input voltage to the rectifier if the DC output is to be satisfactory. Unlike common power supplies that operate with input from a 60 Hz commercial power line, the input voltage to the high voltage rectifier in TV is not highly dependable, and it often fails. That is why the input voltage is checked first. The check for "fire" at the high voltage rectifier anode is a check of the input voltage to be rectified.

Usually the amount of input voltage to the high voltage rectifier is checked by estimating the length of the arc that can be drawn, rather than by measurement with instruments. But the DC voltage at the rectifier cathode should be measured using a voltmeter with a high voltage probe. Sometimes this DC voltage is checked by the arc test, but this is not a recommended procedure because it has the potential for damaging the circuit by permitting an excessive amount of current to flow.

Other than the facts that high voltage is involved and that less filtering capacitance is needed because the frequency is also high, the high voltage rectifier and power supply is a simple, standard basic power supply that follows the same principles as power supplies covered in Chapter 1.

178   The Horizontal Sweep Circuit

In most cases, low or no output from the high voltage DC power supply is the result of low input to the rectifier from the flyback transformer or a defective high voltage rectifier diode. If there is high voltage DC at the picture tube anode but still no raster, the problem must be either a defective picture tube or wrong voltages applied to other parts of the picture tube at the socket terminals.

8-4 Checking for Causes of Loss of Raster Unrelated to High Voltage: The Picture Tube and Electron Gun Voltages

If the high voltage system has been found to be working satisfactorily and there is still no raster, the cause is either a defective picture tube or wrong voltages at the picture tube socket terminals. Picture tubes fail because of open filaments or heaters, shorts between electrodes which make up the electron gun, and low emission. Refer to Table 8-1 for a graphic view of relationships and alternatives.

The best method of checking a picture tube is to use a picture tube checker. This is usually incorporated in an instrument designed to rejuvenate picture tubes. This instrument will test for emission (beam current), shorts in the electron gun, and overall condition of the picture tube. In the rejuvenation mode, the technician can restore satisfactory emission in a large percentage of low emission cases, and can also clear a large percentage of the shorts that develop inside the tube.

If the picture tube is good and high voltage is good, yet there is no raster, what remains to be checked are the voltages applied at the socket terminals to activate the electron gun elements. These voltages should be measured with a DC voltmeter, and values found compared with values specified in service notes. Significant discrepancies should be traced to their cause, and the fault corrected. We should keep in mind that these are DC voltages, usually supplied by the basic low voltage power supply of the receiver. We should look for shorted or leaky capacitors, changed value resistors, and possible shorts or opens in the interconnecting wires or printed circuit board traces.

Occasionally, a pin connection on the picture tube socket will develop an open circuit because the wire from inside the glass is improperly soldered to the socket pin. Hence, the socket and the socket pins should be looked at and checked with extreme care when the trouble persists after all checks previously outlined have been made and all discrepancies from "what should be there" have been traced and their causes corrected. Although rare, it is not unknown in the author's experience for an open circuit to develop, often intermittent, either inside the picture tube in the lead going from the external pins to the electron gun elements or outside the glass in the socket pins as described. Open or intermittent connections to the socket pins outside the glass can, with care, usually be repaired. If the fault lies inside the glass, there is little one can do about it unless a repair can be made with a rejuvenator.

This concludes the first phase of our discussion on the causes of loss of raster, that is, no high voltage, improper voltage on the picture tube socket terminals, or a defective picture tube. Thus far, we have considered only the high voltage power supply itself as a cause of loss of high voltage. There are other causes. The one we examine next is the horizontal amplifier, which drives the high voltage rectifier as a part of its load. A malfunction here also affects the high voltage.

8-5 The Horizontal Amplifier: Functions and Failures

The function of the horizontal amplifier is to accept the signal from the horizontal oscillator and amplify it to provide enough power to move the scanning beam across the picture tube and also to supply input to the high voltage rectifier and power supply.

The output of the horizontal amplifier goes into the primary winding of the horizontal output transformer, popularly referred to as the "flyback transformer." Secondary windings, or taps on the primary winding if it is an autotransformer, provide energy to move the spot across the picture tube, to operate the high voltage rectifier, to feed the damper, to provide pulse voltages to operate the AGC circuit, and to provide other pulse voltages to compare with incoming sync from the transmitter in the AFC circuit to accomplish automatic frequency control of the horizontal oscillator.

One might say that the principal load on the horizontal deflection transformer coils is composed of the yoke, the high voltage power supply, and the damper. A short in any one of these affects all of them. This happens because a short anywhere changes the load on the horizontal amplifier and so affects its output.

The comments made about linearity in analyzing the input signal to the vertical amplifier (Figure 7-3) also apply to horizontal linearity. Horizontal yoke current, too, must rise uniformly during horizontal trace time. Nonlinearity in the horizontal sweep has another dimension, however. In the right-hand side of the picture ($\frac{1}{2}$ to $\frac{2}{3}$, approximately) the spot is moved by current from the horizontal amplifier; on the left-hand side ($\frac{1}{3}$ to $\frac{1}{2}$, approximately) the spot is moved by current from the damper diode. This arrangement puts to use some of the energy released by the collapsing magnetic field and absorbed by the damper as it kills off ringing.

While the details may all be very interesting, the important thing for the troubleshooter to remember is to look to the damper circuit for the cause of horizontal nonlinearity at the left side of the picture, and to the horizontal oscillator and amplifier when the nonlinearity appears at the right-hand side of the picture.

When trouble occurs that affects the horizontal amplifier output it can usually be attributed to one of three categories:

1. The horizontal amplifier tube or transistor has become defective.
2. Either the dynamic input signal from the horizontal oscillator or the static DC voltages applied to the horizontal amplifying device is not right.
3. Something in the load or output circuit of the horizontal amplifier has become defective.

Troubleshooting of the horizontal amplifier requires that we check each of the possible causes of failure, one at a time, until the defective part is found.

We begin our checks with either Category 1 or 2, depending on which is easiest to perform. If the receiver is tube type we start with the first category and check the tube by instrument or substitution. If the receiver is solid state, we will probably begin with Category 2 and check the input signal first; if input signal is proper we next check static conditions (V_C, V_B, V_E), and then the transistor.

If the tube or transistor is good, the signal input is good, and

8-5 The Horizontal Amplifier 181

all static DC voltages applied to the amplifying device are what the service notes call for, we conclude that the problem must lie somewhere in the output or load circuit. Here we have a problem. The load circuit is composed of several parts: the output transformer, the high voltage power supply, the deflection yoke's horizontal coils, the damper, and the minor loads represented by the pulses furnished to operate the keyed AGC circuit and the horizontal AFC circuit. Which do we check? We must consider the order of probabilities: the damper diode breaks down more often than the horizontal output transformer or the deflection yoke, and the minor loads rarely develop a problem that affects horizontal amplifier performance. So usually the damper circuit is checked first. This we discuss in Section 8-6 on the damper circuit.

In checking the flyback transformer and deflection yoke there are three approaches, each with its own advantages and disadvantages. The first approach is to use a flyback transformer and yoke tester, an instrument designed especially for this purpose. This is probably the best method, all factors considered, with the least amount of ambiguity.

The second approach is to use the signal substitution method, which calls for an instrument that will produce an output comparable to that of the horizontal amplifier. The output lead is disconnected from the receiver's horizontal amplifier and the output signal of the instrument is substituted. If the trouble is in the horizontal amplifier itself, this substitution check will produce high voltage and raster. If the trouble is in the transformer or deflection yoke, there will still be improper raster and further checking will be necessary.

The third approach is to use an ohmmeter to check the resistance of all windings of the flyback transformer and deflection yoke and compare these measured values with the values that are specified in service notes. This method will find open circuits dependably but may or may not reveal shorts. If the shorts show up only when the high voltage is present, the ohmmeter will not reveal them.

All methods have some degree of ambiguity that can only be resolved completely by making a judgment and trying a new part in the circuit. The third method is the worst, because the testing voltage of the ohmmeter is so much different from actual voltages present in the circuit when the receiver is operating. The number of bad guesses can be minimized by carefully checking everything for any signs of overheating, burned or discolored insulation, melted wax, brittle wire insulation, signs of previous arcing, lingering odors, discolored resistors, and abnormal sounds when the receiver is turned on.

No matter which method is used to pinpoint troubles in this part of the horizontal sweep system, there will be some wrong conclusions. There is not a television serviceman living who has not at some point in his career concluded that the trouble was in the flyback transformer or the deflection yoke, and installed a new one, only to find that he had not cured the trouble. When this happens, at least one knows that the part that was replaced is good, and he can check other parts by the substitution method as a last resort. This method is dependable, but it takes a lot of time, it carries the possibility of making a mistake in connecting leads of the new part, and it requires a lot of unique parts for substitution, many of which will sit on the shelf for a long time if the fault is found to be elsewhere.

8-6 The Damper Circuit

We have noted that in both vertical and horizontal circuits retrace must be accomplished in a very short time compared to trace time. We have also seen that the rapid collapse of magnetic fields as current falls during retrace induces high voltage pulses and oscillation called ringing. Ringing must be stopped or damped before the next scan begins in order to produce a good picture.

The basic function of the damper circuit is to kill off ringing by the end of the horizontal blanking pulse. To accomplish this the damper diode conducts heavily during that part of the ringing oscillation when the polarity of the ringing voltage constitutes forward bias. The result is rectified voltage (DC) at the cathode of the damper diode. This voltage is called boost or bootstrap voltage in older receivers because it is higher than the voltage of the basic DC power supply. Boost voltage is applied to the horizontal amplifier tube in order to squeeze more power out of it. It may sometimes be applied to other places in the receiver such as the electron gun elements in the picture tube or the vertical sweep circuit. If boost voltage is low, horizontal amplifier output will be low. This causes the high voltage to be low and the sweep to be inadequate, and often leaves a black line along one or both sides of the picture.

With the advent of solid state devices, greater variety is appearing in damping systems. The horizontal sweep circuit is being recognized as a potential source of energy for rectifier circuits that require less capacity for filtering, resulting in lowered costs as well as increased efficiency. Some of these rectifiers are both filtered and regulated, and their output is used to supply DC voltage to operate subsystems that formerly were connected directly to the basic DC power supply.

The damper circuit is a half-wave rectifier. Troubleshooting it is usually a straightforward proposition. Damper troubles commonly produce one or more of the following symptoms:

1. Loss of raster because the trouble disables the horizontal amplifer and the high voltage power supply.
2. The picture fails to fill the screen horizontally.
3. Nonlinearity in the left half of the raster.

The most frequent problem in damper circuits is failure of the damper diode. If the receiver is tube type, the most satisfactory way to check the diode is by substituting a new tube. A tube diode in the damper circuit tends to develop shorts that only show up when high voltage is present. Consequently, the damper will almost always check "good" on a tube checker, because it does not duplicate the high voltages that are present when the tube is in use.

A filter capacitor bypasses the damper cathode in many circuits. This capacitor, subject to high voltage pulses and the boost voltage, sometimes breaks down. A short circuit here is a short across the low voltage DC power supply with the damper in series, and unless the circuit is fuse protected, something will burn out due to excessive current. It is a good idea to check the cathode side of the damper for a short to ground before turning the set back on after replacing any components whose failure might be attributed to excessive current.

Do not attempt to use the oscilloscope to check the damper circuit without first making certain that it is safe and will not damage the oscilloscope. Voltages in the damper circuit exceed the maximum permissible input voltage to the oscilloscope in most cases, unless special precautions are taken.

Nonlinearity in the left half of the raster indicates trouble in the damper, since the damper provides the deflection current for this portion of the horizontal scan. In present day receivers this is not a common problem, but it does occur occasionally. Look for a defective diode or capacitor as the usual cause.

8-7 The Horizontal Oscillator

The horizontal oscillator generates the horizontal sweep frequency of 15,750 Hz. The frequency is usually established by a resonant circuit consisting of inductance and capacitance. The inductance is variable so that the exact frequency required can be set by the technician and/or the owner. On some receivers there is only one horizontal hold adjustment, the resonant circuit inductance. On others, the horizontal hold control is a sort of vernier (usually a variable resistor) which controls only a limited range of frequency. In this case, there is a second adjustment, usually the resonant circuit inductance itself, which provides for wide range adjustment. Ordinarily, this second adjustment is regarded as a "technician's adjustment," and is not primarily intended for the set owner's use.

The output of the horizontal oscillator is acted upon by a pulse-shaping network of resistance, capacitance, and sometimes inductance. Then it goes to the horizontal amplifier as a driving signal for the horizontal sweep.

Troubleshooting of the horizontal oscillator is called for when the driving signal at the input to the horizontal amplifier fails to meet specifications. This, we recall, is one of the major check points; the shaped pulse from the horizontal oscillator is checked on the oscilloscope for proper peak-to-peak magnitude and proper wave shape. Depending on how the oscilloscope display compares with what should be there, the troubleshooter can now decide if the trouble is in the horizontal oscillator or beyond this checkpoint—somewhere in the horizontal amplifier, the high voltage power supply, or the damper circuit. Refer to Figure 8-2 for examples of the waveform at this check point.

The output of the horizontal oscillator may be missing, it may be the wrong frequency, or it may be of improper wave shape. If there is no output from the horizontal oscillator, it usually indicates a catastrophic failure of the oscillator tube or transistor, the DC power source, or some component in the circuit. Check the tube first if tube type, then check static conditions (DC voltages V_C, V_B, V_E) and resistor values; check for shorted or leaking capacitors, and continuity of coils and circuits. If there is still no output and no clues, look for the feedback path and check for an open circuit somewhere in it. Remember, we are dealing with an oscillator which is an amplifier with feedback. Through feedback the oscillator furnishes its own signal; if there is no feedback signal there is no output.

If the frequency is wrong, look first for the frequency-determining elements. Move all adjustable frequency controls through their entire range to see if the circuit is capable of producing the correct frequency. If picture lock-in cannot be achieved, something has gone wrong in the circuit. The next step is to check all oscillator voltages and component values for some clue to what has changed.

If voltages and resistor values all seem normal, a capacitor substitution box or decade box can be very helpful. The box may be connected in parallel (which adds to capacitance already there) or one end of the capacitor in the circuit may be disconnected and the box substituted. Either larger or smaller values can be substituted and their effects on the circuit observed.

It is useful to know if the oscillator frequency is too high or too low. If the blanking bars slope downward to the right the frequency is too high. If the blanking bars slope downward to the left the frequency is too low. See Figure 8-3 for illustration.

Summing up, a change in almost any element of its environment will affect an oscillator's frequency. This includes temperature, DC voltages applied to the tube or transistor, resistance, capacitance, and inductance in the circuit. For frequency control by the viewer, variable resistance or variable inductance is usually used. It is possible that horizontal frequency problems may originate in a fault in the AFC circuit. This circuit produces a DC control voltage to hold the oscillator on frequency and in proper phase. The AFC circuit discussion is presented in the following section.

8-8 Horizontal Automatic Frequency Control (AFC)

As the name suggests, the horizontal AFC circuit controls the frequency of the horizontal oscillator and holds it steady on the correct frequency automatically during the time a picture is being received. When a picture is not being received the horizontal oscillator is not controlled but is free running. This poses no problem, because sync is important only to insure that video information falls at the right point on the raster; when no picture is being received synchronization is unimportant.

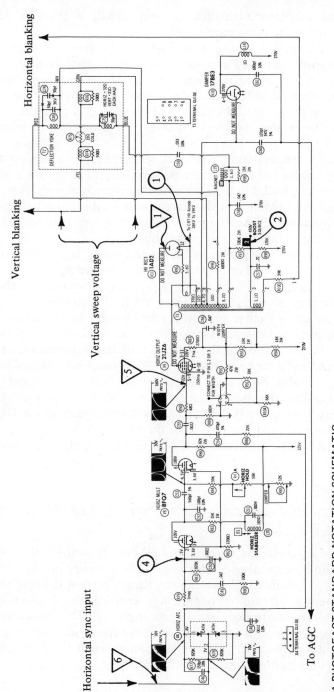

A PHOTOFACT STANDARD NOTATION SCHEMATIC
© Howard W. Sams & Co., Inc. (1969)

(a)

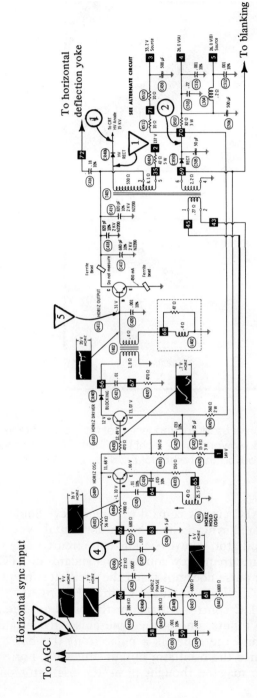

A PHOTOFACT STANDARD NOTATION SCHEMATIC
© Howard W. Sams & Co., Inc. (1974)

Figure 8-2 Typical horizontal sweep systems. (a) Vacuum tube type circuit. (b) Transistor type circuit (courtesy of Howard W. Sams & Co., Inc.).

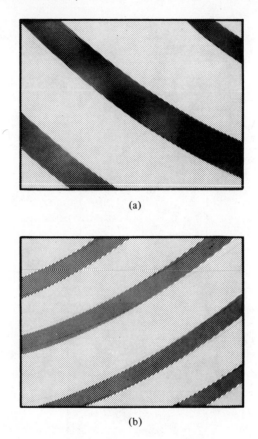

Figure 8-3 (a) Horizontal oscillator frequency set too high. (Note that the blanking bars slope downward to the right.) (b) Horizontal oscillator frequency set too low, causing the horizontal blanking bars to slope downward to the left.

How does the AFC circuit control the frequency? Some early receivers used triggered sync, in which the horizontal sync pulse itself timed the oscillator. However, current circuit design has settled upon automatic frequency control as standard. In AFC circuits today the sync pulse does not reach the oscillator; frequency is controlled by a *DC voltage* produced by the AFC circuit.

To produce the DC control voltage for the oscillator the AFC circuit requires two inputs: one is the sync pulse that originates in the transmitter and the other is a pulse from the horizontal amplifier output. Since the oscillator output drives the horizontal amplifier, this pulse may be thought of as the oscillator output

8-8 Horizontal Automatic Frequency Control (AFC)

signal, except that it is amplified and isolated by the horizontal amplifier. These two pulses are fed into the AFC circuit, which compares them and from the comparison derives a DC voltage.

The sync pulse is fixed in frequency; the output of the oscillator is variable. When the oscillator is "on frequency" a certain DC control voltage exists. If the oscillator tends to drift higher in frequency the control voltage changes in one direction; if the oscillator tends to drift lower in frequency the control voltage changes in the other direction. This action may be observed by measuring the control voltage with a DC voltmeter (with the picture locked normally) while we vary the horizontal hold control first one way and then the other. Try this on a properly working receiver. It is a good point to remember what should happen when AFC problems arise.

Key elements in AFC circuit operation are:

1. Two input pulses must be present, one must be the sync pulse, the other must be derived from the output of the horizontal oscillator at some point.

2. The AFC circuit must compare these two pulses and produce a compensating DC control voltage for the oscillator whenever the oscillator tends to drift out of phase and off frequency.

3. The free-running frequency of the horizontal oscillator must be set close to the correct sweep frequency of 15,750 Hz. This is done by the horizontal frequency adjustment of the oscillator. Then the AFC circuit will take over to make it exactly right as long as the oscillator is close to 15,750 Hz and the AFC circuit is working.

Troubleshooting AFC circuits begins with the oscilloscope. We must verify that both sync and horizontal pulses are present at the input to the AFC circuit. If they are not present, or if they are of the wrong values, the circuit will not work properly. If both pulses are present at the inputs to the AFC circuit, we next measure the DC control voltage as we raise and lower the horizontal frequency. The DC control voltage should vary as frequency moves above and below 15,750 Hz. If the control voltage does not change, check the diodes and other parts of the AFC circuit. Very often, a breakdown of one of the diodes will prevent the circuit from working properly.

Should it not be clear whether the trouble is in the AFC circuit or the horizontal oscillator, one technique is to remove the control voltage by opening (or sometimes grounding) the circuit. This leaves the oscillator free-running. Now, carefully adjust the horizontal frequency as you watch the picture. If the picture can be almost stabilized, but drifts easily, it indicates that the horizontal oscillator is working and that its frequency can be adjusted to be proper but it lacks a proper control voltage. This suggests that the trouble is in the AFC circuit. If the picture cannot be adjusted to be almost stable, the problem is in the oscillator.

8-9 Summary

The horizontal sweep system is a multifunction subsystem in television receivers. It provides horizontal sweep, high voltage, pulses for AGC, pulses for AFC, and incorporates a damping circuit that produces useful output voltage. All of these functions are interrelated and, for the most part, interdependent.

Troubles in the horizontal sweep system produce visible symptoms. This places a premium on careful observation and thoughtful analysis of what is seen by the troubleshooter. Special care must be taken when dealing with high voltages present in this system. They are dangerous to people and equipment. High voltages also tend to break down insulation and circuit elements, resulting in problems that are sometimes ambiguous to standard test instruments.

Most horizontal problems are straightforward and will yield to the techniques we have discussed. Some problems, especially in color receivers, present some interesting twists that are a challenge to anyone. Here, perseverance and careful, thoughtful analysis pays off. Remember, theory *does* apply; if a circuit seems to be behaving contradictory to theory it is because you are overlooking something.

chapter 9

Color television:
Elementary servicing procedures

In 1954, the Federal Communications Commission (FCC) adopted standards for the present system of color television in the United States. Approximately ten years passed before color television became a significant factor in the marketplace. The system that was adopted is a compatible system, which means that a color signal transmission can be received as a black and white picture on a monochrome receiver, and a monochrome signal transmission can be received as a black and white picture on a color receiver.

To incorporate color information with the brightness information already included in the monochrome signal, certain modifications in signal standards and receivers were required. Basically, a subcarrier of approximately 3.58 MHz is modulated on the picture carrier. Color hue and intensity values are carried by the subcarrier. Circuits must be added to the receiver to process the subcarrier and recover the color hue and intensity values. And, of course, the picture tube in the receiver must accept color information in the form of signal voltages, and from them produce a color picture.

In other words, the color receiver is much like the monochrome receiver; the major differences are in the picture tube and in the circuits added to deal with the color subcarrier. Thus, basic troubleshooting techniques that apply to monochrome TV receivers also apply to color receivers. The new dimensions in color TV servicing are the direct result of new circuits that had to be added to the receiver to handle the added characteristics of color.

It is not our objective here to go deeply into the technical details of color television or to treat color receiver troubles exhaustively. (Complete books have been written on both subjects.) Rather, the purpose of this chapter is to present practical procedures for attacking some of the more common problems of color TV, problems that do not require the knowledge or experience of an expert troubleshooter to solve.

9-1 Similarities and Differences in Monochrome and Color Television Receivers

Because the color system is compatible with monochrome, there must be similarities, but adding color must make for some differences. Here at the outset, let us point out some of the specific similarities:

1. Similar basic DC power supplies are used in both receivers.
2. Both receivers are superheterodynes, with similar tuners, IF amplifiers, and video detectors.
3. Sound is similar in both receivers.
4. Both receivers have similar AGC and horizontal AFC control circuits.
5. Vertical and horizontal sweep circuits are basically similar in both receivers.
6. Servicing techniques and procedures for circuits thus far listed are similar.

It is also interesting to note some of the ways in which color and monochrome receivers differ. Virtually all differences are located after the video detector. (Refer to Figure 5-1 and Figure 9-1 for block diagrams):

1. The color picture tube differs in several details from the monochrome picture tube even though they are both CRTs, both use electron guns to form their beams, both require high voltage, and both use electromagnetic deflection.
2. There is an amplifier for the color subcarrier, or chroma, (3.58 MHz) that is not found in the black and white receiver.
3. There is usually an amplifier for the color burst, or color sync, signal.
4. There is a crystal oscillator that is controlled by the color burst.
5. There are color (chroma) demodulators.

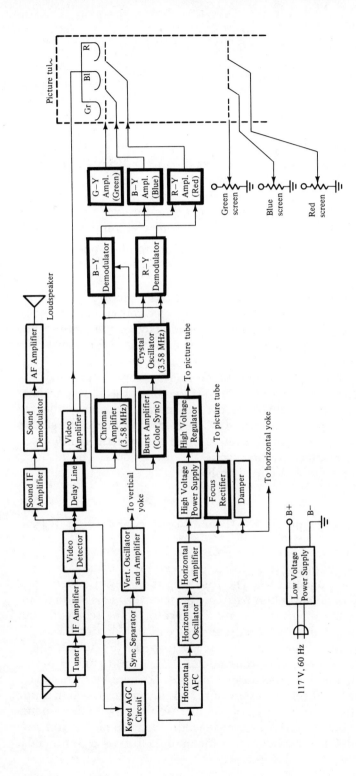

Figure 9-1 Functional block diagram of basic color television receiver. Note that heavy lines call attention to new, added circuits required to process the color signal. Refer to Figure 5-1 to compare with monochrome receiver.

6. The high voltage supply is usually regulated in the color receiver.
7. There is usually a separate focus rectifier in the color receiver.
8. There are three sets of color signal amplifiers in the color receiver in addition to the video (brightness) amplifier which drives all three cathodes; one set drives the red gun of the picture tube, one set drives the green gun, and the third set drives the blue gun. The monochrome receiver has only one channel for video amplification and one electron gun.
9. There are new circuits and adjustments to obtain proper color and convergence.

Let's begin our discussion of how a color picture is produced with a brief look at the traditional three electron gun picture tube in Figure 9-2. The phosphor on the inside face of the picture tube consists of a fine mosaic of dots of chemically different makeup that glow red, green, and blue when struck by an electron beam with sufficient energy. When the system is properly adjusted, the electron beam from the blue gun strikes only the blue phosphor dots, the electron beam from the red gun strikes only the red phosphor dots, and the electron beam from the green gun strikes only the green phosphor dots. Should the adjustment be improper, the electron beams may not strike the proper dots. The result is what is referred to as a purity problem.

9-2 Purity Adjustment in Three Gun Color Picture Tubes

The question of purity is a primary one, and it is usually resolved prior to other adjustments. The first step is to degauss (demagnetize) the chassis and metal parts of the receiver. This is done automatically by a built-in circuit in current color receivers, but must be done manually by a degaussing coil in older models.

Following degaussing we are ready to begin the purity adjustment sequence. Normally, only one of the electron guns is turned on at a time for this procedure. This may be done by turning two of the screen controls to minimum and advancing the remaining screen control until the face of the picture tube lights up, producing a red, blue, or green raster, depending on which control was advanced. Refer to service notes for manufacturer's recommenda-

9-2 Purity Adjustment in Three Gun Color Picture Tubes 195

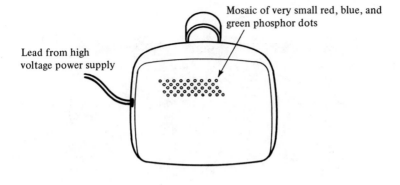

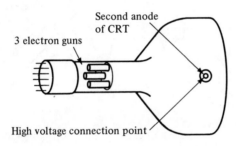

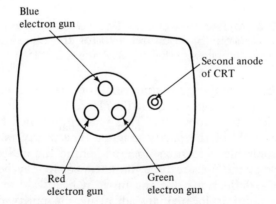

Figure 9-2 Basic structural details of the common three electron gun type of color picture tube.

tions relating to specific instructions and the location of controls for a particular receiver.

If the raster produced by a single electron gun is not a pure and uniform color throughout, it means that some of the beam

196   Color Television: Elementary Servicing Procedures

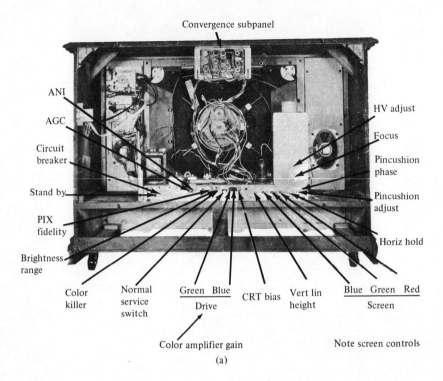

Figure 9-3 (a) Rear view of color TV receiver showing physical location of various controls. (b) Expanded detail of controls and locations in the same receiver (courtesy of Howard W. Sams & Co., Inc.).

from that particular gun is not striking the color dot of the phosphor that it is supposed to strike. Here our problem is to direct the beam to the proper dot. This is accomplished by varying the magnetic field of the purity magnet. The purity magnet consists of two rotatable, thin rings, located behind the deflection yoke on the neck of the picture tube. Sometimes, it may be necessary to loosen the deflection yoke and move it backward or forward until the best point is located. By adjusting the purity magnet a uniformly colored screen should be obtained with no off-color or impure areas. When all three guns have been checked we are ready for the next step. This is to adjust for a white screen. See Figures 9-3 and 9-4 for location of the various controls and the purity magnet.

If the brightness of the red, blue, and green phosphor dots is properly balanced, and if the viewer is far enough away that his

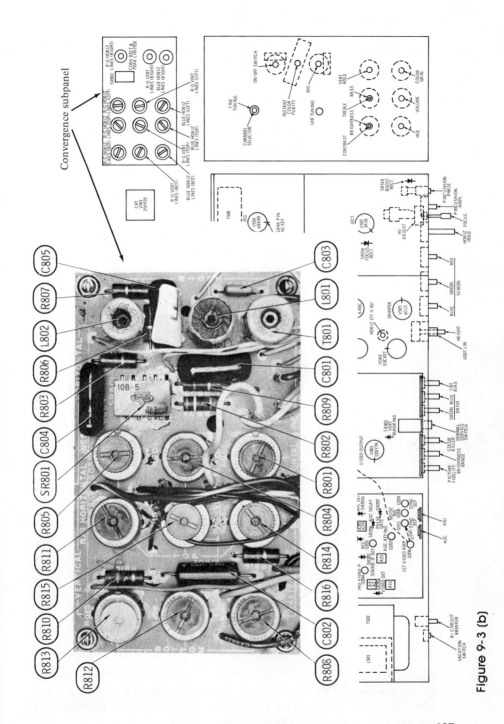

Figure 9-3 (b)

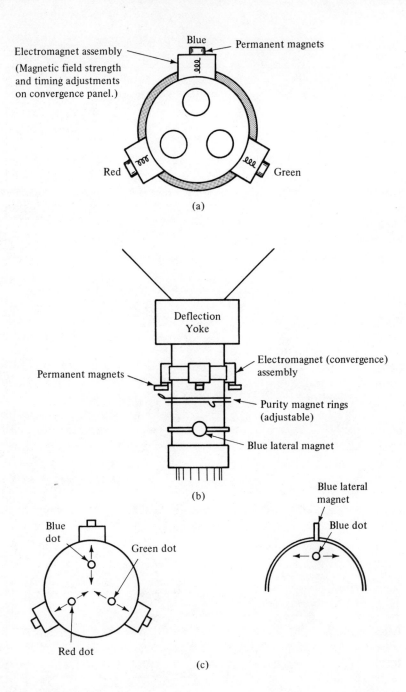

Figure 9-4 (a) Physical location and orientation of convergence assembly on the neck of the picture tube. (b) Location of various essential elements as viewed from above picture tube. (c) Direction of movement of lighted spots available to achieve convergence.

eye cannot perceive the individual color dots, the screen will appear white. To make this adjustment the brightness control is set near maximum and all screen controls are turned to minimum. This should make the screen dark; if not, the kinescope (picture tube) bias control should be adjusted, or the brightness control if there is no bias control, until the screen is dark. Now, the red screen control is advanced until the screen is barely lighted with a uniform red. Next, the green screen control is advanced until it changes the color on the screen. Finally, the remaining blue screen control is advanced until it mixes with the red and green. By experimentally adjusting the blue and green screen controls without touching the red, a white screen can be produced when the proportions are right.

Most receivers have a second set of controls which control the gain of the amplifiers that drive the electron guns of the color picture tube. These controls are adjusted to maintain the white screen at different brightness levels within a picture when the color control is set at minimum. Stated another way, the different color gain controls, in conjunction with the screen controls, are adjusted to maintain the proper color balance required to produce white throughout the grey scale on a monochrome picture.

The technician should refer to service notes on the particular receiver in question to obtain step-by-step instructions for both purity adjustments and grey scale adjustments. These are manufacturer's recommendations and, while they are not inviolable, they do represent the manufacturer's thinking as to what best fits his particular design. With practice, the technician will develop his own sequences and short cuts and will find that he will have to refer to service notes only in cases where there are differences in design.

9-3 Convergence

After adjusting the purity and the grey scale for white, the next problem is to adjust for proper *convergence*. To explain what this term means we observe that each electron gun receives the same brightness information as does the single electron gun of a monochrome receiver. Hence, each gun produces a picture, one in red, one in green, and one in blue. When properly converged, each picture falls precisely on top of the other two, and the viewer sees only a single picture. If convergence is not proper, the viewer will see color fringes because the separate color pictures are *out of registry* and do not fall exactly on top of one another. The process of going through the adjustments required to make the pictures coincide exactly and to eliminate color fringing is referred to as converging the picture, or converging the receiver.

How do we converge a receiver? What instruments are needed, what adjustments must we make, and are there any special problems we should be alerted to watch for?

First, we should understand that convergence is achieved by manipulating magnetic fields. Both permanent magnets and electromagnets are used. The permanent magnets are usually mounted on the electromagnets, which are placed around the neck of the picture tube behind the deflection yoke and directly over the electron guns. This is shown in Figures 9-3 and 9-4.

The only instrument required for convergence is a dot-bar generator. Using this, the operator is able to select a signal which will produce a dot pattern on the screen or, if he chooses, a crosshatch pattern. Usually, the generator signal is connected to the antenna terminals and the tuner is set to channel 3 or 4 to pick it up. The generator and the receiver are adjusted to produce a clear, stable dot pattern on the picture tube. If the receiver is properly converged, the dots produced by the red, blue, and green guns fall *exactly* on top of each other and the viewer sees only a complete field of white dots across the entire face of the picture tube. If the receiver is not properly converged, some or all of the dots do not fall exactly on top of each other, and the viewer sees color dots.

When convergence adjustment is required, it is recommended that we converge dots in the center of the screen first, disregarding dots elsewhere. To do this, the permanent magnets, which shift the entire field of a particular color, are adjusted to converge the center dots. The red and green guns can be shifted in only one dimension, as shown in Figure 9-4. The blue beam, on the other hand, is provided with a second permanent magnet, called the blue lateral adjustment magnet, in addition to the one that is similar to the red and green adjustments. These two magnets enable the blue field to be moved in two dimensions at right angles to each other. This geometry is necessary to provide enough flexibility to achieve convergence. The blue beam can be moved vertically by the permanent magnet on the electromagnet assembly and horizontally by the blue lateral magnet. The blue lateral magnet is located behind the other magnet assembly, over the blue gun on the neck of the picture tube.

We can see that by moving the red and green fields (and dots) as shown in Figure 9-4c, at some point the two dots will converge. It is also clear that by simply moving the blue field (and dots) up and down may not enable blue convergence with the already merged red and green dots; horizontal movement is necessary to give the required flexibility. This is the function of the blue lateral magnet.

So, after shifting the entire color fields first to converge the center dots, we must look very critically at the rest of the screen. *Misconvergence away from the center calls for adjustment of the electromagnet assembly.* These controls are often grouped together on a subpanel not normally accessible to the set owner. The subpanel convergence controls affect the electromagnets. They control the magnitude and phase, or timing, of the current pulses in the three electromagnets, which in turn control the red, blue and green electron beams at the top and bottom, and the left and right margins of the picture. Refer to Figure 9-3 for subpanel location and designation of convergence controls found on it.

At this point, a definite sequence of adjustments is preferable. Until one has much experience it is best to refer to service notes and follow the sequence that is presented there. The example we have chosen is taken from Howard W. Sams' *Photofact* Set 1124, Folder 2, dated 9-70, on Packard-Bell color chassis 98C21. We have designated this Figure 9-5.

Purity and convergence adjustments are referred to as set-up adjustments. They are performed prior to putting a color receiver into service. After that, they are made only when major repair or other factors make them necessary. When all set-up adjustments are completed (AGC, vertical height and linearity, focus, etc.) the receiver is ready for use, assuming there is no circuit problem or breakdown that might interfere.

9-4 Problems in Color Television Receivers

If we compare the block diagrams and signal paths in monochrome and color TV receivers, we see that problems fall largely into two categories, one color-related, and the other noncolor-related.

In the color receiver the noncolor-related problems are the same as for black and white receivers. This means that troubles in the tuner, the IF amplifiers, the video detector, the sound section, the vertical sweep section, the horizontal sweep section, sync and AGC are identical in both monochrome and color re-

CONVERGENCE ADJUSTMENTS

Step	Control	Use to Converge (or Straighten)	Remarks
1.			Perform Center Dot Convergence using convergence magnets. See Fig. A.
2.	R-G Vertical Lines, Top R815	Red and Green Vertical bars at top of screen.	Touch up both controls for best convergence from top to bottom along vertical center line (Fig. B).
3.	R-G Vertical Lines, Bottom R812	Red and Green Vertical bars at bottom of screen.	
4.	R-G Horizontal Lines, Top R811	Red and Green Horizontal bars at top of screen.	Touch up both controls for best convergence of horizontal bars along vertical center line (Fig. B).
5.	R-C Horizontal Lines, Bottom R813	Red and Green Horizontal bars at bottom of screen.	
6.	Blue Horizontal Lines, Top R814	Blue Horizontal bars at top of screen.	Touch up both controls for best convergence of horizontal bars along vertical center line (Fig. C).
7.	Blue Horizontal Lines, Bottom R808	Blue Horizontal bars at bottom of screen.	
8.			Perform Center Dot Static Convergence (Fig. A).
9.	Blue Horizontal Lines, Right T801	Blue Horizontal bars at right side of screen.	Touch up both controls for best convergence along horizontal center line (Fig. D).
10.	Blue Horizontal Lines, Left R801	Blue Horizontal bars at left side of screen.	
11.	R-G Vertical Lines, Right L801	Red and Green Vertical bars at right side of screen.	(Fig. E)
12.	R-G Horizontal Lines, Right L802	Red and Green Horizontal bars at right side of screen.	Use control to converge blue bar with red and green bars on right side of screen (Fig. E).
13.	R-G Vertical Lines Left R804	Red and Green Vertical bars at left side of screen.	(Fig. E)
14.	R-G Horizontal Lines, Left R805	Red and Green Horizontal bars at left side of screen.	Use control to converge blue bar with red and green bars at left side of screen (Fig. E).

ceivers. Symptoms and troubleshooting techniques are also identical. Material covered in Chapters 5 through 8 apply equally to the above mentioned sections of the color receiver.

Let us, then, consider the most common color problems, their

9-4 Problems in Color Television Receivers   203

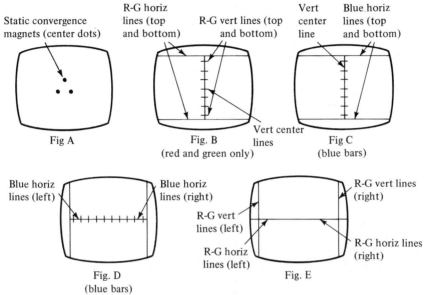

Figure 9-5 Convergence instructions (left) found in service notes for Packard-Bell Chassis 98C21 (courtesy of Howard W. Sams & Co., Inc.). Note use of illustrations (above) to clarify written sequence of adjustment of specific controls.

symptoms, and what to do to correct them. In Table 9-1 we present color problems, using the same format as we used in Table 5-1 when we considered monochrome receiver problems. Again, we divide the problems into two groups, troubles due to improper adjustment and troubles due to circuit breakdown. Some, as before, can result from either maladjustment or circuit breakdown.

In troubleshooting color receivers the first question to be answered is whether the trouble is a color problem or not. Does the receiver produce a good black and white picture, and is the sound normal? Do tuning controls, and brightness, contrast, vertical and horizontal hold controls operate normally? If the receiver has AFT (automatic fine tuning) does the set operate properly on manual as well as AFT? Once the question of whether the problem is a color problem or not has been decided, the next step is to investigate the circuits concerned.

When problems in color TV overlap, the trouble affects both color circuits and other circuits in some way. Power supply problems, for example, are often in this category. In general, whenever it is difficult to discern if the problem is strictly a color problem or a noncolor problem, it is best to attack and resolve the non-

TABLE 9-1a

Color TV Receiver Troubles That May Often Be Corrected by Adjustment

Adjustment Problem	Cause and Location	Remedy
1. Wrong color in some areas of the picture.	Purity improper; adjustments wrong or metal parts and mounting brackets have become magnetized.	Adjust purity magnet; if necessary, degauss receiver and adjust deflection yoke. Follow procedure of Section 9-2.
2. Figures in picture have colored outlines.	Convergence improper; permanent magnets on convergence assembly and/or electro-magnets require adjustment.	Perform sequence of convergence adjustments. Refer to Section 9-3 and Fig. 9-5.
3. People's faces are too red or too green.	Hue control improperly adjusted.	Adjust hue control for proper flesh tones.
4. No color in picture.	Fine tuning wrong; color control set too low; color killer set improperly.	Adjust fine tuning, color control, and color killer as required.
5. Stripes of color appear like a rainbow across the picture.	Crystal oscillator is off frequency.	Check adjustments that affect crystal oscillator frequency. (This oscillator has limited frequency range.)
6. Color is washed out and weak.	Brightness level set too high and/or color control set too low.	Adjust brightness level and color control as necessary to produce a good color picture.
7. Color is good on some stations, but there is no color on others.	Local oscillator in tuner not adjusted properly; color killer adjustment not set right.	Adjust local oscillator in tuner (Chapter 4, Section 4-2). Adjust color killer as necessary.
8. One color seems to dominate; colors not true	Electron guns in picture tube are not properly balanced.	Recheck set-up adjustments to produce a white screen and proper grey scale. Refer to Section 9-2.

TABLE 9-1b

Common Color TV Receiver Troubles, Symptoms, Locations, Checkpoints and Procedures

Trouble	Probable Good Sections	Possible Trouble Locations	Procedures, Checkpoints, and References
1. No color on any channel; black and white picture good.	All except those incorporated especially to handle the color signal.	a. Chroma amplifier b. Crystal oscillator c. Other color circuits	Use oscilloscope as dynamic signal tracer to localize fault. Then make static checks of circuit to pinpoint the defect.
2. Monochrome pictures have color tint.	All except grey scale adjustments and circuits, and possibly the picture tube.	a. Voltages on electron gun elements not proper b. An electron gun in the picture tube is defective	Check grey scale adjustments, check static voltages on electron guns. Check the color picture tube.
3. Color stripes across picture.	All but burst amplifier and crystal oscillator.	a. Color burst is not reaching crystal oscillator b. Crystal oscillator is off frequency	Use oscilloscope to check for burst signal and for output from crystal oscillator. Make static checks to see what went wrong when trouble has been localized. Repair as needed.
4. One color is apparently missing in the picture.	All except color amplifier color demodulators, static voltages on electron guns, and the picture tube.	a. Fault in color amplifier b. Defect in color demodulator c. dc voltages on electron guns are wrong value	Check grey scale adjustments first. Make static checks on color amplifier circuits. Check dc voltages on electron guns, check the color picture tube.
5. Focus poor; may change at different brightness levels.	All except high voltage power supply and focus circuit.	a. High voltage regulator is bad b. Fault in HV power supply c. Focus rectifier circuit is defective	Use VTVM and high voltage probe to check output of high voltage power supply, high voltage regulator, and the focus rectifier. Repair as necessary and recheck for proper operation.
6. Color comes and goes, yet the monochrome picture is steady.	All except the circuits that handle only color.	a. Crystal oscillator is intermittent b. Chroma amplifier bad c. Other color stages are defective	Use oscilloscope to determine if crystal oscillator is intermittent, or if chroma signal is reaching demodulators. Make static checks to pinpoint trouble.

206 Color Television: Elementary Servicing Procedures

TABLE 9-2a

Flow Chart for Troubleshooting Color TV
Problem Number 1

Problem: Picture poor, no color on any channel.
Possible causes: 1. Trouble in dc power supply.
2. Trouble in noncolor circuits.
3. Trouble in color circuits.

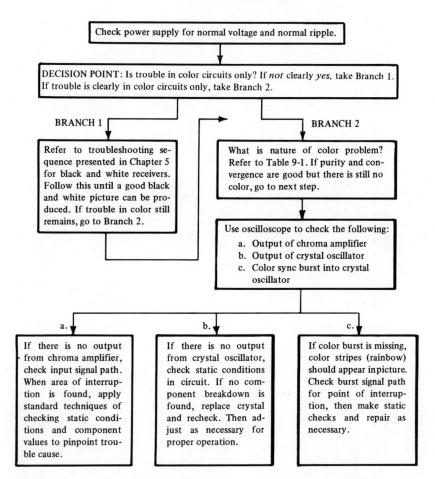

color problem first. As soon as it is assured that the receiver will produce a good black and white picture and the noncolor controls function normally, then we can proceed to troubleshoot the color section. A flow chart suggesting a sequence of steps is presented in Table 9-2.

TABLE 9-2b

Flow Chart for Troubleshooting Color TV Problem Number 2

Problem: Convergence, black and white picture and sound good, but cannot achieve proper color balance.

Possible causes:
1. One of the color amplifiers has improper output.
2. There is trouble in one of the demodulators.
3. A picture tube electron gun is defective.

---

1. Check monochrome picture carefully. If it is good, the picture tube is probably good. If it is not good and shows poor grey scale, i.e., tends to show color instead of shades of grey, check the picture tube. If all electron guns check normal, look for trouble in the color amplifiers driving circuitry or in the circuits used to reach proper grey scale at all brightness levels.

---

2. Determine which electron gun is causing the imbalance. Check dc voltages, starting with the picture tube cathode and grid bias voltages. Compare measured voltages with service notes specifications of what should be there. If a discrepancy is found, continue investigation until the cause is found and corrected. If all seems to conform to specifications, proceed to next step.

---

3. Check for proper static conditions in the demodulator circuits. Trace the cause of any discrepancies noted and repair and correct problems as needed.

The procedure just described will resolve the majority of color imbalance problems. For those cases where trouble still persists it is necessary to use a full dynamic testing procedure in which we inject the signal from a color bar generator and proceed to check the gain and dynamic operation of each stage until the cause of faulty operation is discovered and corrected.

To properly handle frequencies found in color TV, such as the 3.58 MHz color burst, an oscilloscope capable of 5 MHz response (minimum) is recommended. If the oscilloscope is not capable of resolving the color burst signal, the troubleshooter is handicapped by not being able to determine with confidence if the signal is present or not.

Troubleshooting color receivers, as in monochrome, is a process of knowing the signal paths, making dynamic checks of signal progression down each path, and checking the static conditions to find out what went wrong when a circuit fails to perform dynamically as it is supposed to.

The principles involved in many of the color circuits are not really new. For example, the chroma amplifier is much like an IF stage that is tuned to handle 3.58 MHz. The crystal oscillator, like other electronic oscillators, turns DC from the power supply

TABLE 9-3

FCC Standards for Color TV Transmission in the United States

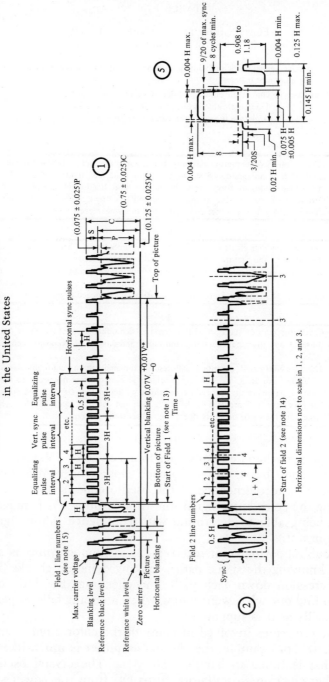

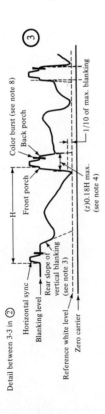
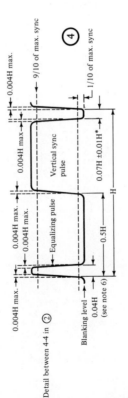

1. H = Time from start of one line to start of next line.
2. V = Time from start of one field to start of next field.
3. Leading and trailing edges of vertical blanking should be complete in less than $0.1H$.
4. Leading and trailing slopes of horizontal blanking must be steep enough to preserve minimum and maximum values of $(x + y)$ and (z) under all conditions of picture content.
*5. Dimensions marked with asterisk indicate that tolerances given are permitted only for long time variations and not for successive cycles.
6. Equalizing pulse area shall be between 0.45 and 0.5 of area of a horizontal sync pulse.
7. Color burst follows each horizontal pulse, but is omitted following the equalizing pulses and during the broad vertical pulses.
8. Color bursts to be omitted during monochrome transmission.
9. The burst frequency shall be 3.579545 MHz. The tolerance on the frequency shall be ±10 Hz with a maximum rate of change of frequency not to exceed 1/10 Hz/second.
10. The horizontal scanning frequency shall be 2/455 times the burst frequency.
11. The dimensions specified for the burst determine the times of starting and stopping the burst, but not its phase. The color burst consists of amplitude modulation of a continuous sine wave.
12. Dimension "P" represents the peak excursion of the luminance signal from blanking level, but does not include the chrominance signal. Dimension "S" is the sync amplitude above blanking level. Dimension "C" is the peak carrier amplitude.
13. Start of Field 1 is defined by a whole line between first equalizing pulse and preceding H sync pulses.
14. Start of Field 2 is defined by a half line between first equalizing pulse and preceding H sync pulses.
15. Field 1 line numbers start with first equalizing pulse in Field 1.
16. Field 2 line numbers start with second equalizing pulse in Field 2.
17. Refer to text for further explanations and tolerances.

into AC (3.58 MHz). It can also be described as an amplifier with feedback that provides its own signal. Unlike many oscillators, it is not free running; it is locked in frequency and phase by the 3.58 MHz color burst signal sent out from the transmitter. Only a small adjustment is provided by the hue control, by which the viewer can adjust the phase to match that of the original carrier so that a proper color balance can be recovered in the demodulators.

The troubleshooter should keep in mind that, in all stages, the dynamic mission to perform first requires that the proper static conditions be established and maintained. Color TV circuits share this in common with all electronic circuits, regardless of the complexity of the theory, the supporting mathematics, or the size and purpose of the total system. The thoughtful, logical individual with a qualitative understanding of circuit behavior can deal effectively with many practical electronics problems even though he may possess surprisingly little command of the supporting theoretical structure.

chapter 10

Troubleshooting safety, general procedures, and special problems

This chapter is written directly to you, the reader. It includes topics not covered elsewhere or only touched upon lightly. You might say this chapter is a response to the fact that we do not live in an ideal world. Things are not always safe, you and I do not have perfect memories, and some problems we encounter are very fleeting and elusive.

There are many facets that contribute to successful troubleshooting, including paying attention to safety, establishing good interpersonal relationships, having access to adequate service notes and sources of replacement parts, and keeping good records. It also helps if you are a good practical psychologist. Notice that we have yet to mention command of theoretical knowledge and technical expertise, choice and use of instruments, experience, and ability to deal effectively with special problems.

Remember that a repairman is always dealing with people with problems. That is why they come to you. And, because they are people with problems, they are often likely to be a bit testy, impatient, dour of countenance, and sometimes downright unreasonable. This is understandable, if you think about it. How do you feel when you find you must take your automobile to the shop, wondering about costs, about how long your car may be unavailable to you, and perhaps about the competence and integrity of the person who will do the work? The situations you will be working in call for special understanding and patience. Be prepared to spend a little extra time to put your customer at ease and to explain clearly the options as honestly as you possibly can. Be careful that you do not underestimate the problem. Above all do not make firm promises that you may not be able to carry out.

10-1 Safety

Safety has two dimensions, the safety of individuals and the protection of equipment from damage. Safety of people is the number one priority in all electronics work. This is complicated by the fact that electricity is invisible and great danger may lurk in an innocent-looking wire. Good safety policy calls for constant awareness of hazards and continuous vigilance to avoid sloppy practices.

Current magnitudes of 100 milliamperes (and sometimes less) through the human body are potentially lethal. The question is, how much voltage does it take to cause a current of this value? Or more to the point, what is your resistance? According to Ohm's Law, if your resistance is 5,000 ohms it would require 500 volts of electrical pressure to produce 100 mA, and if your resistance is 1000 ohms it would require only 100 volts to produce 100 mA. In some surgical techniques, electrodes that make contact directly with the blood encounter resistance on the order of 50 ohms. Now, only 5 volts of electrical pressure are required to produce 100 mA, so even a small voltage is potentially lethal.

The conclusion is plain: be respectful of voltages in excess of 30 volts. You will live longer, have fewer scars, and have a much better disposition to present to your grandchildren.

When working with low voltage transistor circuits you may tend to be lulled into a sense of false security. This can make you careless and fall into bad practices. Carelessness is especially dangerous when you are called upon to repair vacuum tube circuits; these rascals love high voltages of 100 volts and above, and they have no respect at all for the "mellow" transistor worker. So, be doubly careful when working with vacuum tube circuits!

The hazard of working with transformerless power supplies is discussed in detail in Chapter 1. One way to avoid this hazard is to always operate the equipment you are working on from an isolation transformer. Otherwise there is always the possibility of inadvertently placing a direct short across the AC power line. This is potentially dangerous to both technician and equipment.

Equipment damage from improper use has two aspects: (1) The test instrument may be damaged by too high voltages and currents, especially pulse voltages such as those found in TV sweep circuits; (2) An improperly used low sensitivity DC voltmeter may cause damage in a high gain directly coupled transistor amplifier under test by upsetting critical bias voltages the moment the meter leads are connected. The solution to Item 1 is to know the maximum safe input voltage to each instrument you use and be careful that it is never exceeded. The solution to Item 2 is to use a high input resistance voltmeter, such as a VTVM or DVM, which will not upset the circuit being measured.

A final word about safety may surprise you. It concerns housekeeping. Good housekeeping around your work area is important. I have personally seen cabinets and plastic safety masks so deeply scratched by a screw carelessly left lying on top of the work bench that they had to be replaced. I have seen plastic parts melted and deformed, cabinets burned, and service notes charred by carelessly placed soldering irons. I have seen carelessly dropped items stepped on and damaged beyond repair. We are all subject to human frailties, and all of these things will happen to all of us at some time. But the vast majority of these "accidents" can be avoided by the practice of good housekeeping.

There are other reasons besides safety for practicing good housekeeping. Good housekeeping saves time, because you have a place for things and you return everything to its place when it is not being used. This way, you know where to find a tool or other item when you need it. Let's make hunting for tools and misplaced parts a thing of the past!

The discipline that you must apply to maintain good housekeeping will become a part of you. This is valuable in itself. Don't forget that in electronics you are working with invisible forces, and a disciplined approach is more productive, less frustrating and much, much safer.

10-2 Following a Job Through

There is a general sequence of steps to be followed for every repair job that comes to you. It is, of course, subject to modification to fit special circumstances. The general sequence is presented in Table 10-1. We will look at various parts of it that call for more explanation and elaboration.

Please note especially one of the preliminary steps titled "History of the Trouble." The value of this step is often underestimated. Much troubleshooting time can be saved by taking a moment to find out if the trouble occurred suddenly, indicating a catastrophic failure of some part of the circuit, or if the trouble developed slowly over a long period of time. A good history that

TABLE 10-1

General Troubleshooting Flow Chart

START: You assume responsibility for the repair of an item of equipment.

PRELIMINARY STEPS: *Make out job card to include the following:*
a. Name, address, and telephone number of equipment owner.
b. History of the trouble.
c. Any promises made relating to costs, time, delivery, etc.

Verify the trouble:
a. Check the equipment to see if it performs as stated.
b. Test all switches and operating controls if the trouble is not clearly evident.
c. Assess reported history with symptoms you see. Make an initial estimate of the nature of the trouble.

DO INITIAL TROUBLESHOOTING: *Remove access covers but do not disassemble further.*
a. Check fuses, cords, switches, controls, vacuum tubes.
b. Look critically for signs of arcing, overheating, and shorted connections. Use sense of touch (heat), smell (odor), sight, and hearing as you search for anything unusual that might constitute a clue.
c. Review all that has been done so far. Decide if you know enough to make repair now or if further detailed trouble shooting will be required.

PERFORM DETAILED TROUBLESHOOTING: *If detailed troubleshooting must be done, do the following:*
a. Assemble service notes and tools.
b. Collect and set up instruments.
c. Remove chassis as necessary to gain access to test points. Connect extension leads as required to make unit operable.

Collect experimental data and compare with what should be there.
a. Measure dc voltage and ripple at output of the power supply. Refer to Chapter 1 if values are not normal.
b. Choose signal tracing or signal injection to localize trouble to the defective stage or subsystem.
c. When defective stage or area is found, make checks of static (dc) voltages, circuits, and component values.
d. Repair any breakdowns found.

TEST REPAIRED UNIT FOR PROPER OPERATION:
a. Apply power, check unit for proper operation.
b. If all seems well, test run unit for a long enough time to be certain it reaches stable operating temperature.
c. If unit operates properly, reassemble, and test run once again.

COMPLETE PAPERWORK:
a. Itemize work done, parts replaced, time spent, and costs.
b. Replace tools, finish housekeeping details, file your records and service notes.
c. Call owner, deliver, or fulfill whatever arrangements were made when you took responsibility for the job.

→ THE JOB IS NOW COMPLETE!

is thoughtfully taken and analyzed can frequently be a direct clue to the trouble.

If you know that the trouble is intermittent before you start the troubleshooting process, how will this affect the way you approach the problem? Will you do the same things in the same sequence as you would if you knew the equipment had smoked and then went dead? Or if you discover that several other technicians before you have attempted to locate and repair the problem without success, how optimistic will you permit yourself to be when giving an estimate of time and repair costs to the equipment owner? And if the history reveals that there is intermittent snow in the TV picture which becomes worse when the wind blows, how much time will you devote to searching for trouble in the receiver before you test out the antenna system?

The history of the trouble will frequently suggest where to start looking for the trouble. With this in mind, the next step is to verify the owner's complaint. You must see for yourself exactly how the equipment malfunctions. Symptoms seen by someone who is not technically knowledgeable may be described incompletely and in a different way than you or I would describe them.

From this point you start the initial troubleshooting procedure. Actually these steps merge and overlap, rather than exist as clear-cut, separate phases as it may appear when we describe them. You add each new bit of information to the whole as you go along. Troubleshooting is much like assembling the pieces of a jigsaw puzzle; you put the data together piece by piece until suddenly the whole thing falls into place.

In the initial troubleshooting procedure, the back of the equipment and other access covers are removed. Tubes, fuses, and other accessible parts are checked and replaced as necessary. In this step you do everything possible to locate the trouble without removing the chassis or doing other major disassembly.

If the initial troubleshooting process fails to identify the trouble you have to remove the chassis. You must make whatever disassembly is required to reach needed test points, circuits, and components that were not accessible before.

At this point (if you have not done it earlier) it is advisable to get and look over service notes pertaining to the equipment. A good set of service notes includes any special disassembly instructions that should be observed. Now, you follow through the general troubleshooting procedure of the flow chart presented in Table 10-1.

The most effective method of attack is to localize the trouble to the faulty stage or subsystem as rapidly as possible. It is a good idea to check the output of the DC power supply at the start, since a fault here will affect the rest of the unit and interfere with proper operation. Once a power supply is exonerated as the cause of the trouble, use either the signal injection or signal tracing techniques, or both, to test subsystems for proper dynamic operation.

Every piece of equipment possesses strategic test points that are valuable for making quick checks. Examples are shown in Figure 2-3 for radio and Figure 5-1 for television receivers. After all, dynamic checking is operational checking, performed to verify if a stage or subsystem is capable of handling a signal as specifications call for.

Any failure other than improper tuning of resonant circuits is always the result of improper static conditions. That is, a circuit has become open or shorted, a component has changed its value or characteristics, or improper voltages have been applied.

As soon as the trouble has been localized, you next make a step-by-step check of static conditions. Usually this begins with DC voltage checks. You save time by beginning here since most circuit and component troubles reveal themselves by upsetting DC voltages. When a wrong DC voltage is found, you check off all the possible causes of this condition one at a time until the cause is pinpointed. Next, replace or repair parts as necessary, then recheck and test the unit for proper operation.

Every unit should be test run long enough for you to feel assured that you have indeed located the trouble and corrected it, or to indicate if further investigation is necessary. This should be done before reassembly. If operation is normal, the unit is reassembled and test run once again. (Sometimes a problem may be introduced in reassembly, or the temperature differences inside the cabinet may reveal a problem that was missed before.) If everything operates normally, complete your paper work and prepare to return the unit to normal use.

10-3 Special Problems and Techniques

Of all the problems that arise in electronic gear, intermittent operation can be among the most difficult to resolve. Intermittents take several forms, some mechanical in nature, some due to changes in temperature, some resulting from changes in voltages or other factors ranging from vibration to dust in the environment.

Vacuum tubes are particularly subject to intermittent ills, usually mechanical. The primary technique to find these is to gently tap the vacuum tube while the equipment is operating; listen and watch closely for any disturbance in sound or picture as the tube is jarred. Tapping and gently bending or stressing components, leads, and printed circuit boards constitute some of the most effective ways to cause a mechanical intermittent to show up.

A major problem in dealing with intermittents is that the unit, in many cases, operates normally most of the time. When the equipment is operating normally you can find no problem because there is no problem; only when the trouble occurs can it be found. Therefore, before you can find the cause of an intermittent either you must make it show up somehow, or you must wait for it to appear spontaneously.

Good practice says that you should wait for the trouble to appear spontaneously *only* after all conscious efforts to cause it to appear have failed. Conscious efforts mean those things that the troubleshooter does to make the problem occur at his command. The tapping, bending, and stressing described earlier falls into this category. Another procedure that will be helpful in some cases is to operate the unit for a time at a lower voltage, say 105-110 volts, and then a higher voltage, say 120-123 volts, before applying the tapping and stressing technique.

Sometimes a crack in a printed circuit board or trace may be invisible to the eye, and efforts to make it show up reveal only the general area of the trouble. When this is the case a repair can frequently be made by the following method. Take a small tipped, low wattage soldering iron and drag it along each trace in the trouble area, reheating (or adding) solder all along each trace. In this way any hairline crack will be repaired, even though you never saw the crack and will probably never know its exact location. This is the frustrating part. So too is the fact that you can conclude you have been successful in repairing the unit only after a long period of operation during which time it remains stable even when tapped and stressed.

Intermittent operation resulting from changes in temperature as the unit warms up brings its own special challenges and frustrations. The first challenge often arises when the chassis is removed from the cabinet to make test points accessible. Now, with air circulating more freely the temperature does not build up as far or as fast as inside the cabinet and the trouble may refuse to show up. If it doesn't show up, you can't find it. That is your dilemma. What are the alternatives in this situation?

The first hurdle in solving the problem is how to raise the heat. Here you have a choice from several possibilities. You can put a cover over all or part of the unit while you watch for the trouble to appear. Or you might try directing a heat source such as a heat lamp or heat gun at the suspect area. The heat gun is a device similar to a hair dryer that is made for such applications.

When these weapons are brought to bear, heat-related problems can usually be made to reveal themselves. However, the troubleshooter may be able to identify only the general area, not to pinpoint the specific spot or component. Here, another weapon is very helpful, namely, spray-on refrigerant. This is commercially available in aerosol cans. With it, small areas and components can be selectively cooled as you monitor equipment operation. With both heating and cooling at your command, the cause of most heat-related intermittent problems can be identified within a reasonable length of time.

When a unit has been in use for a long time without service there is usually a dust buildup that may reach amazing proportions. Often the dust is accompanied by a small amount of settled-out grease that holds the dust firmly. The net result is a covering that interferes with ventilation and contributes to greater than normal heat buildup. Sometimes dust provides a leakage path for current, or it causes arcing. Removal of dust by blowing it out by compressed air or by vacuuming it out should be a definite part of every repair job.

10-4 Instruments, Tools, and Miscellaneous Service Aids

Servicing of electronic equipment, like any other type of service work, requires a certain basic complement of equipment. Certain instruments are used over and over and form the backbone of data gathering essential to analysis. In fact, for our purposes we might classify instruments under the following general headings:

1. *Instruments used for data gathering or data acquisition.* This includes ohmmeters, AC and DC voltmeters, an oscilloscope, vacuum tube and transistor checkers, inductance- and capaci-

tance-measuring instruments, frequency counters, ammeters, and wattmeters.

2. *Signal sources.* This includes an audio frequency sine and square wave generator, a radio frequency generator that is amplitude modulated to cover the regular broadcast band and any other bands to be worked with, a sweep (FM) generator to cover the FM radio band, a TV alignment generator, a dot-bar generator, a color-bar generator, and a flyback tester.

3. *Power sources.* This includes an isolation transformer, DC power supply with adjustable output that is preferably voltage regulated and with adjustable current limiting incorporated in it, a bias box and, perhaps, a battery eliminator for 12 volt systems.

4. *Miscellaneous instruments and tools.* This classification includes such items as a noise generator, a heat gun, accessory probes for instruments, a soldering iron with different wattage tips, a vacuum cleaner, an air compressor, vacuum tube extension sockets, extension leads, alignment tools, a socket set, a screwdriver set, an allen wrench set, diagonal cutters, long-nose pliers, slip-joint pliers, wire strippers and a resin core solder.

5. *Chemical aids.* The following are available and recommended in aerosol spray cans: contact cleaner, refrigerant, insulating film, lubricant, and quick setting adhesive. Other useful items are rubber cement, epoxy or other strong glue, furniture and scratch-cover polish, and a good glass cleaner.

6. *Library.* The extent of your library will depend on your own personal tastes and how much work you do that requires reference material. Library resources should include good service notes, cross reference of parts and components that can substitute for each other, equipment and supply catalogs, tube and transistor manuals and data books, information and service notes on your test equipment, application notes from manufacturers, names and telephone numbers of suppliers and other resource people, and anything else you feel will be helpful.

The person who is acquiring his first test instruments will be confused by the competing claims of various manufacturers. He sees before him different combinations of instruments ranging from top laboratory grade to kits for the hobbyist.

A frequently asked question is, "What about kits?" Any answer to this question must be carefully qualified. In terms of quality, the components of most kits are comparable to those of average-cost test instruments that are manufactured and marketed for servicing consumer products. A more-to-the-point question is concerned with quality control of the completed unit. That is, the physical aspects of mounting the parts, soldering, placement of leads, and so on, must be compared. A carefully built kit can be fully equal to the commercially built unit in these respects if care is taken.

A very important point is the question of calibration and testing to assure that the completed instrument fulfills its specifications. For simple units such as voltmeters this is not a great problem. But for more complex units such as oscilloscopes, the kit builder seldom has either the expertise or facilities for verifying all of their calibration and whether or not they fully meet specifications.

Kit building takes time. Depending on what value you place on your time, the amount of money saved will vary if you count your time as part of the cost. Usually, if time is added, the total cost of the kit will be about equal to its commercially available counterpart. If you have plenty of time, if you enjoy building circuits, and if you feel you are equal to the construction, testing, and evaluation challenges, kits provide a way to obtain a useful instrument for fewer dollars.

Unless you are going to be doing a significant amount of technically demanding, advanced servicing of top quality high fidelity or similar equipment, it is hard to justify the cost of laboratory grade equipment, such as made by Techtronix. Many companies such as B & K, Hickok, and Sencore make a wide variety of field tested instruments designed expressly for the electronics service industry. The list of manufacturers is long.

Be aware that there are various approaches to instrumentation. You must think about size, weight, portability, dimensions, ease of use, space available, and durability as well as initial costs. There are compact units available that include a cluster of allied instruments to accomplish certain functions or procedures from beginning to end, or to accomplish various functions related to a particular type of equipment, such as color television, for example. This may be preferable in some situations to having a separate instrument for each of several special needs. Some shops construct an instrument panel, which may be mounted on a movable cart, or may be built in and stationary. Do you prefer to bring the instruments to the work, or the work to the instruments?

10-5 Parts and Inventory

A question faced by every service organization, large and small, is what operating philosophy to follow with regard to inventory. Should a large inventory of supplies and replacement parts be kept on hand so that you will be able to repair almost all work immediately from stock, or should a minimum inventory of only the most in demand items be maintained, and others ordered as the need arises?

The problem just described has several facets. To carry a large inventory requires a large capital investment. Also, many items will prove to be slow moving. This means that the return on money invested is very low and may be far in the future after taxes, shelf space, inflation, and the fact that your capital is tied up are figured in. On the other hand, being able to do the job immediately attracts customers and represents a competitive advantage. If you are not close to your supply sources, ordering and waiting for parts can take much time.

Many shops which are close to their suppliers may be able to get supplies within hours. Usually, these shops carry a minimum inventory. In a sense, they use and live from their suppliers' inventory. This has the advantage of freeing their capital for other, possibly more productive uses.

No matter how inventory is handled, you should establish good relations with your suppliers. Your supplier is in a unique position to call to your attention special price breaks and manufacturers' promotions as they arise, and he usually has considerable discretionary area within which he can operate. Be considerate of him and his needs and he will be considerate of yours. A good working relationship here can be an asset of great value.

10-6 Troubleshooting Equipment Containing Integrated Circuits

Just as the transistor has supplanted vacuum tubes in circuits made up of separate, discrete components, the integrated circuit

(IC) is now poised to supplant entire circuits involving discrete components. The IC is composed of an entire functional circuit, or circuits, constructed on a single small chip of silicon. The result is an entire amplifier, signal source, or logic gate in a small package that requires only that you provide DC power and signal input in order for it to function. None of the individual elements (resistors, transistors, interconnections) that make up the circuits on the chip can be seen or identified from outside the package.

Now, you can buy complete audio amplifiers, IF strips, demodulators, and a variety of other functional circuits as off-the-shelf items. Connection pins on the chip provide for application of the DC voltage required to operate, and for signal input and output connections. Other pins permit external elements to be connected for feedback, frequency compensation, stability, gain control, and flexibility.

Troubleshooting ICs calls for the same basic techniques that we have discussed for circuits that are composed of discrete components. First, essential static (DC) conditions must be established as a first priority; otherwise, it is impossible to achieve proper dynamic operation.

The first step in troubleshooting is to measure the DC voltages at the device pins to see if measured values agree with "what should be there." If the measured voltages are not right, the cause of the discrepancy must be found and corrected before proceeding further. A wrong DC voltage on *any* kind of device has three possible causes:

1. Something is wrong in the circuit supplying the DC voltage to the device.

2. Something is wrong with the device itself.

3. Some of the other pins of the device have wrong DC voltages on them which upset other voltages in the circuit.

If only a single pin of an IC has an improper DC voltage on it you must make a series of voltage checks, following the circuit back to its source point at the output of the DC power supply. If the source voltage is normal, and nothing along the circuit carrying the voltage to the device is wrong, the IC is defective. Replace it. If the circuit still does not work, you have either overlooked a fault in the circuit you have just examined, or there is a defect (an open or short) in the circuit carrying the signal to or from the device, or the new device is bad.

If more than one pin has wrong DC voltage on it, there is a

good chance that just one of the improper applied voltages is the fault which upsets the essential static conditions, and the other wrong voltages are the device's normal response to the upsetting voltage. In this case, you must isolate the device from the circuit, either by removing the device from its socket or by disconnecting the device pins where wrong voltages were found. Now, with the device isolated, each circuit must be checked carefully to see if voltages are what they should be when the device is not connected. If all circuits are proper, the device is defective. Replace it. If the same wrong voltages are still present, you have overlooked a circuit fault. Go back and recheck; the fault is there somewhere!

Dynamic testing of ICs is no different from dynamic testing of discrete component circuits. You turn power on, apply the proper signal, (magnitude, frequency, waveshape) at the device input and then check for the proper signal at the device output pin. *If all static conditions are right, it has to work!*

Troubleshooting circuits containing ICs should not be feared. ICs are typically mounted on PC boards, usually with smaller, closer together traces than are found on PC boards made for discrete component circuits. This means that IC troubleshooting requires more emphasis on isolating the IC to enable checking for circuit faults on the PC board and its connection points.

For ICs, the causes of system malfunctions form a different profile of frequency than for discrete component circuits; ICs show a higher percentage of total failures in circuits and in devices. That is, short circuits, open circuits, and ICs themselves constitute a higher percentage of the total number of failures. Indeed, how could it be otherwise? Nearly all of the resistors and transistors are gone as discrete, separate parts, vanished into the IC package. So the number of external parts subject to failure in a particular circuit is much fewer than before; hence, the percentage of total failures is altered.

The work of electronics servicing and troubleshooting is like a game, full of challenges, but open and amenable to solution by the skillful and resourceful. It can be very rewarding to anyone who gains personal satisfaction from problem solving, more so because the solved problem is a real, tangible piece of equipment restored to usefulness. If you like to see immediate, tangible worthwhile results from what you do, try electronics troubleshooting. It's a growing field!

Index

A

AC circuits, 12
AC current, 12
Alignment, intermediate
 frequency, 90-99
Alternating current, 13
Alternator, 17
Ammeters, 11
Amplifiers, 17, 64:
 Audio, 55-78
 Horizontal, 179-82
 Intermediate frequency,
  79, 83, 84-90
 Push-pull, 70-76
 Radio frequency, 79-100,
  101-12
Aspect ratio, 139
Automatic gain control,
 139-60, 155-57

B

Batteries, 17
Boxes, substitution, 29
Bridge-rectifier power supplies,
 43-44

C

Capacitance, 13-15
Capacitor checkers, 29
Capacitors, 13-15, 52
Circuits, 8-11, 16-19
 AGC, 157-60
 Damper, 182-83
 Horizontal sweep, 173-90
 Integrated, 221-23
 Ohm's law applied to, 12
 Parallel, 34
 Performance, 7
 Short, 34-36
 Vertical sweep, 161-72
Closed circuit, 9
Closed series circuit, 10
Complete circuit, 9
Control elements, 15-21
Conventional current, 12
Convergence, 199-201
Current, 8, 12-13

D

DC circuits, 12
DC current, 12

226 Index

Diode checkers, 29
Diodes, 34, 43-44

E

Electric field, 14
Electromagnetic field, 14
Electron flow, 11-13
Electronics systems, 15-16
Electrons, 12
Electrostatic field, 14
Energy sources, 15-21
Experimental data, 7

F

FETVOM (Field-Effect-
    Transistor-Volt-Ohm-
    Meter), 11
Federal Communications
    Commission, 191
Field-Effect-Transistor-Volt-
    Ohm-Meter
    (FETVOM), 11
Fields, 14
Frequency, 13
Fuel cells, 17

G

Generator, 17
Ground points, 50-53

H

Horizontal automatic
    frequency control,
    185-90

Horizontal sync pulse, 142-44

I

Inductance, 13-15
Inductors, 13-15
Instruments, 11

L

Loads, 15-21
"Lumps," 13, 14

M

Magnetic field, 14
Mathematical predictability,
    11-13
Microammeter, 11
Milliammeter, 11, 29
Mixers, 101-12

O

Ohmmeter, 11, 29, 34, 77
Ohm's law, 11-13
Open series circuit, 10
Oscillators, 101-12
    Horizontal, 184-85
Oscilloscope, 29, 64, 143

P

Parallel circuit, 10, 16, 34
Persistence, 142
Power distribution systems,
    15-16
Power supply, 27-53

Preselectors, 101-12
Purity adjustment, 194-99

R

Raster, 139, 178-79
Receivers, television, 113-37, 139-60
Rectifier, 30, 37-43
Recycling, 2, 4
Resistors, 13-15
Resistance, 13-15

S

Safety, 212-13
Scanning, 139
Series circuit, 9-10, 16
Series-parallel circuit, 10-11, 16
Short circuits, 34-36
Signal generators, 82-83
Signal injection, 63-65
Signal tracing, 63-65
Solar cell, 17
Substitution boxes, 29
Subsystems, 21-25
Superheterodyne, 101-5
Sync pulses, 142-53
    Vertical, 154-55
Synchronization, 139-60

T

Television, color, servicing of, 191-210
Television receivers, 113-37, 139-60
Thermocouple, 17
Transducer, 17, 19

Transformers, 28, 29, 30, 40-43
    Isolation, 37, 38
Transistors, 76-78
Tube checkers, 29
Tuners, 108-11

V

VOM (Volt-Ohm-Milliammeter), 11
VTVM (Vacuum-Tube-Volt-Meter), 11
Vacuum-Tube-Volt-Meter (VTVM), 11
Vacuum tubes, 72
Vertical sync pulse, 154-55
Volt-Ohm-Milliammeter (VOM), 11
Voltage, 31
Voltage doublers, 44-45
Voltage measurements, 65-70
Voltage regulation, 45-50
Voltmeters, 11, 51
Voltage, 178-79

W

Wattmeter, 29
Waveforms, 53